JOHN GLOAG ON INDUSTRIAL DESIGN

Volume 5

DESIGN IN MODERN LIFE

DESIGN IN MODERN LIFE

Edited by
JOHN GLOAG

LONDON AND NEW YORK

First published in 1934. Second impression 1946 by George Allen & Unwin Ltd.

This edition first published in 2023
by Routledge
4 Park Square, Milton Park, Abingdon, Oxon OX14 4RN

and by Routledge
605 Third Avenue, New York, NY 10158

Routledge is an imprint of the Taylor & Francis Group, an informa business

© 1934, 1946 John Gloag

All rights reserved. No part of this book may be reprinted or reproduced or utilised in any form or by any electronic, mechanical, or other means, now known or hereafter invented, including photocopying and recording, or in any information storage or retrieval system, without permission in writing from the publishers.

Trademark notice: Product or corporate names may be trademarks or registered trademarks, and are used only for identification and explanation without intent to infringe.

British Library Cataloguing in Publication Data
A catalogue record for this book is available from the British Library

ISBN: 978-1-032-36309-7 (Set)
ISBN: 978-1-032-36611-1 (Volume 5) (hbk)
ISBN: 978-1-032-36617-3 (Volume 5) (pbk)
ISBN: 978-1-003-33292-3 (Volume 5) (ebk)

DOI: 10.1201/9781003332923

Publisher's Note
The publisher has gone to great lengths to ensure the quality of this reprint but points out that some imperfections in the original copies may be apparent.

Disclaimer
The publisher has made every effort to trace copyright holders and would welcome correspondence from those they have been unable to trace.

DESIGN IN MODERN LIFE

Edited by

JOHN GLOAG, Hon.A.R.I.B.A.

With contributions by
ROBERT ATKINSON, F.R.I.B.A., ELIZABETH DENBY, Hon.A.R.I.B.A., MAXWELL FRY, B.Arch., F.R.I.B.A., JAMES LAVER, FRANK PICK, Hon.A.R.I.B.A., A. B. READ, A.R.C.A., R.D.I., and GORDON RUSSELL, R.D.I.

LONDON
GEORGE ALLEN & UNWIN LIMITED
MUSEUM STREET

FIRST PUBLISHED IN 1934
SECOND IMPRESSION 1946

All rights reserved
PRINTED IN GREAT BRITAIN BY
BRADFORD & DICKENS, LONDON, W.C.I.

INTRODUCTION TO THE SECOND EDITION
DESIGN IN LIFE AND INDUSTRY, 1934-1944

by

JOHN GLOAG

SINCE this book first appeared in 1934, a new and largely unacknowledged industrial revolution has started. Before 1939 this revolution had suggested a hopeful release for mankind from all manner of hampering old-world habits of design. New and promising materials were gradually coming to the service of industry. The place of the industrial designer, long recognized in the United States and Europe, was at last securing some recognition in the country that had once led the world in industrial production.

The commercial machine age in which Europeans and Americans had been living for over a century seemed to be moving inexorably towards the provision of universal abundance, in goods, services and leisure. The Mistress Art, Architecture, was inspired with new life in the Western world, and was emerging from the phase of crude functionalism. Things built during that phase had repelled ordinary people by a puritanical severity of form. But preoccupation with functionalism was a necessary and healthy reaction after the flamboyant period of over-ornamentation, of "applied art," that had afflicted nearly every branch of design from the early years of the nineteenth century to the end of the first world war, a period which had produced things as variously ridiculous as the Tower Bridge with its stone trimmings, and brass tea trays impressed with a lizard skin pattern. Unfortunately, the second world war interrupted the development of the new industrial revolution. The modern movement in design was arrested in 1939; its achievements, so far as Europe was concerned, were petrified. Only in America was it carried forward for a short time. In the nineteen thirties the United States had been enriched by many splendid experiments in design, industry and large-scale regional planning. The most conspicuous example of the latter was the Tennessee Valley Authority scheme, which created not only magnificent machine architecture, but housing schemes, civic and industrial buildings, upon a scale hitherto unmatched outside the U.S.S.R. In the provision of housing for war workers, American industrialists and designers have been able to solve many problems of design compactly, economically and speedily during the war years; and until Pearl Harbour, American industrial design and development were largely dedicated to satisfying the vast market represented by the forty-eight states of the Union. So the U.S.A., has advanced further than Britain in applying the new materials and methods produced by the new industrial revolution. Across the Atlantic, the modern movement in design is not a case of arrested development; but there is a real danger that it may become so on this side, for its practitioners, generally engaged in war work, do not appear to have progressed; they seem to have stayed put mentally in their outlook on civilian problems of design, if their occasional pronouncements and writings are any indication of what they really think.

Do they still stand for stern and uncompromising functionalism?

Do they realize that at home the civilian population has been subjected for years to rationing, necessarily severe and cheerfully endured because the wartime need of it was recognized?

Do they realize that compulsory "utility" designs in such articles as cups and furniture have been shorn of every trimming, every humanizing touch of ornament, that would have endeared them to ordinary people?

Do they realize how heartily sick everybody is of plainness and austerity, imposed by the state?

Do they imagine that "utility" designs have educated the taste of

INTRODUCTION TO THE SECOND EDITION

the public, so that henceforth people will accept plainness and austerity with thankful enthusiasm, when it is dished out to them as "the modern movement in design"?

The beliefs of the modern designer before 1939 had an alluring simplicity: he, or she, had a faith which supplied all the answers; only, of course, it wasn't called a *faith*. It was called something high-sounding and intimidating like "the logical adjustment of economic facts and mechanical means to functional ends," and it masqueraded as reason, pure-burning and undimmed by any concessions to ordinary human likes and dislikes. Will such beliefs persist unmodified?

In *Architecture Arising*, a book published in the autumn of 1944, Howard Robertson says that "the reforms in architecture which the contemporary movement in design has contributed—and they are of lasting importance—risk being compromised by the narrow and uncompromising outlook which the reformer ultimately assumes towards all who do not toe the line. History proves that no one can be more inhuman than the successful reformer, for the time comes when he is positively infuriated by nonconformity in others."*

Now another book on architecture, of comparable significance with Howard Robertson's work, appeared in 1944, from the pen of Maxwell Fry, entitled *Fine Building*. Its author contributes Chapters II and IX to the present volume. It will be instructive to quote some penetrating criticisms of this book, made by a reviewer in *The Architectural Association Journal*, because they deal fairly and concisely with recent architectural history in this country; and to understand what is happening in contemporary architecture, is an essential preliminary to understanding what is happening, and what could happen, in nearly every branch of design that affects everyday life. The reviewer said:

"Most of us, if we had set out to write a book on architecture in the past four or five years, would have written very much this kind of book—though I doubt if we should have written it so well. It portrays very accurately what

* Chapter I, page 23.

was going on in the minds of many 'younger' architects at the end of the peace. It is, if you like, the 'modernist' story—the story about the Good Fairy of steel and glass and concrete and the Bad Ogre of slums and land-rackets and 'period' façades. It is a true story; but, somehow or other, reading it again, and looking at the slick shadow-conscious photographs, I found myself rather less enthusiastic about this story than I was five, ten, or fifteen years ago.

"Thinking it over, I discovered that Fry's unbounded confidence in his story had acted as an irritant and a challenge. 'Of course, I know you're right,' I found myself saying, 'but is it really as sublimely true as all that? I rather doubt it?' And I began to think of some of the awkward things about 'modern' architecture. As, for instance, the fact that few people outside the MARS Group and a few hundred loyal partisans among the laity really like it—even after fifteen years of strenuous propaganda. Fry does not say much about this but it is rather important, because architecture is more and more concerned with masses of people and less and less with sophisticated individuals. Even admitting that the masses are incorrigibly stupid (and I doubt Fry's humanity would allow him to make such an admission), it is going to be rather difficult, in a democratic State, to plant on them an architecture which they so cordially dislike. And arising from this, I found myself wanting to ask Fry whether there were not perfectly valid reasons why the ordinary man is so singularly indifferent to the charms of the 'modern.' Whether, for instance beauties arising from a three-dimensional, 'organically conceived' design are not, inevitably, obscured from anybody except the architect and those who have the opportunity of seeing the published plans and isometrics (the modern architect thinks isometrically; the layman does not). Whether, in fact, modern architecture is not as obscure and, to most people, inaccessible, as some modern poetry.

"I believe there are answers to these difficulties, but Fry, borne along by his own creative enthusiasm, by happy analogies with physiology and machinery and by a number of very beautiful pictures, does not attempt to supply them. But the moment is coming when they will have to be answered. If modern architecture is to be really objective—and that is its claim—it must bring the aesthetic needs of the man in the street (not in an airplane or at a drawing-board) within the range of its objectivity. If he is going to accept

INTRODUCTION TO THE SECOND EDITION

architecture and all its claims, he must have an architecture whose language appeals to him from street level. Of course, there is only one possible answer—and that is in the form of architecture in the solid. It will come. But meanwhile it is possible that the 'modern' story has its shortcomings. It may be nothing but the truth; but it is not the whole truth."*

It may well be asked: is the "modern" design story just another lullaby for our critical faculties? This extensive quotation from the Journal of one of the most important and influential architectural bodies in Britain not only shows the perplexities that confront the student of design, but suggests that his critical faculties must be unsleeping. We have a duty to doubt the convenient and tidy creeds that, like some patent medicines, are said to cure or at least to alleviate practically every ill. Good design is not a matter of faith. It is easy, but wholly misleading, to say: "This must be right because it is functional." "Fitness for Purpose," the old slogan of the Design and Industries Association, gets you nowhere unless you remember the second part of it, which is: "and pleasantness in use." An artifact, whether it be a building, a teapot, a ship or an electric iron, must be pleasant in use. Fitness is not enough. Functionalism is not enough. Unlit by imagination, functionalism and fitness may end in the grimness of *inhumanism*. The modern movement in design could easily have degenerated into that dark opposite of liberty and variety in life.

There has been an unfortunate and confusing tendency among highbrow critics, writers and some designers, to identify methods of design with political beliefs. Although unacknowledged by its dupes, this tendency received continuous support from pre-war German propaganda, which condemned what it called "the art of the Left." All over Europe, practitioners of the modern movement played into the hands of Dr. Goebbels, by proudly but foolishly asserting that their movement expressed the progressive idealism of Left Wing politics. This identification was not only absurd, but largely untrue.

* July-August 1944. Vol. LIX No. 688, page 97.

They would hardly have claimed that the vast industrial projects carried out by Raymond Loewy for various American concerns—such as the Pennsylvania Railroad—could thus be classified. But generally the reforming modernists of Europe and Britain preferred to ignore the United States. They had one point of view only: their own. The other chap was always wrong.

Now there are generally far more than two sides to every question. Persistent and mischievous attempts to identify every human activity with some particular brand of political faith are calculated to undermine the critical faculties of the English race, and this danger which has developed over a long period, was strikingly condemned by Winston Churchill in a speech he made in 1937, when he said:

"There is one aspect of publishing activity upon which I think we should cast a close scrutiny. I mean the deliberate publication of books of a uniform political tendency to an organized mass of readers.

"I do not care whether it is the Left or Right side of politics to which this process is directed.

"To have an elaborate process set up to feed a particular kind of leaf to a particular tribe of injurious caterpillars, incapable of taking any other nourishment, and who take their colour as well as their food from the foliage on which they crawl, is entirely contrary to the spirit of literature or the means of disseminating knowledge.

"Nothing can be worse than to introduce totalitarianism into the field of literature, and to try to breed in a single country races of men and women incapable of understanding one another.

"The glory of literature in any free country is its variety, and the most fertile means from which happiness may be derived in life is from variety.

"The issues in the world to-day are such that readers should be on their guard against any attempt to warp their intellects or to narrow or enfeeble their judgment by tendencious literature with facts increasingly coloured and statistics ever more carefully selected.

INTRODUCTION TO THE SECOND EDITION

"The ordinary man and woman in the age in which we move has to have a new vigilance and to be alive to new perils, to see that they are not being sucked in by propaganda.

"We see whole nations the slave of propaganda, we see great States in which only one opinion is to be tolerated, which is contrary to the genius of man and the inherent urge of human nature."*

At the end of this book there is a list of works concerned with design. This list suggests a course of reading, planned to cover various aspects of design in life. The books in it are recommended because they put a variety of views before the reader. Some of their authors believe that there is only one way of solving any problem; some of them over-simplify the complex subjects they discuss; a few manage to suggest either with sorrow or enthusiasm that all problems of design are insoluble until education, which is only now beginning to affect vast masses of people, has in four or five generations completed its civilizing task. Draw your own conclusions, but remember that for several years Europe has been corrupted by the belief that there are only two sides to every question. In Britain we know better, and that is one of the reasons why, if we retain and use our gifts for tolerance and compromise, we may find that in our country the modern movement in design becomes gracious and human, and that our second industrial revolution is a triumph, and not, like the first, an ugly disaster.

JOHN GLOAG,
October, 1944.

* Speech delivered at the opening of the *Sunday Times* Book Fair, 1937.

CONTENTS

INTRODUCTION

CHAPTER I
WHO KNOWS WHAT THE PUBLIC WANTS? ... 17
By John Gloag

CHAPTER II
THE DESIGN OF DWELLINGS ... 29
By E. Maxwell Fry, B.Arch., A.R.I.B.A.

CHAPTER III
THE LIVING-ROOM AND FURNITURE DESIGN ... 39
By Gordon Russell

CHAPTER IV
CLOTHES—AND DESIGN ... 49
By James Laver

CHAPTER V
DESIGN IN THE KITCHEN ... 61
By Elizabeth Denby

CHAPTER VI
THE DESIGN OF ILLUMINATION ... 73
By A. B. Read, A.R.C.A.

CHAPTER VII
DESIGN IN PUBLIC BUILDINGS ... 83
By Robert Atkinson, F.R.I.B.A.

CHAPTER VIII
THE DESIGN OF THE STREET ... 97
By Frank Pick, President, Design and Industries Association

CHAPTER IX
DESIGN IN THE COUNTRYSIDE AND THE TOWN ... 111
By E. Maxwell Fry, B.Arch., A.R.I.B.A.

CHAPTER X
THE MEANING AND PURPOSE OF DESIGN ... 125
By Frank Pick

A LIST OF BOOKS ON DESIGN ... 135

LIST OF ILLUSTRATIONS

Between pages

THE DESIGN OF DWELLINGS — 30 and 31
- Plate 1. The elementary cottage type
- ,, 2. The house of the social townsman
- ,, 3. Formal street architecture, and slum street architecture
- ,, 4. Suburban street architecture (speculative picturesque)
- ,, 5. New multiple-flat dwelling
- ,, 6. Flat dwellings for the poor. (Füchsenfeldhof, Vienna.) And fairy palaces for tired business men (Anywhere, England)
- ,, 7. Modern materials and the sort of house they can make
- ,, 8. The free interior of a modern house

THE LIVING-ROOM AND FURNITURE DESIGN — 38 and 39
- Plate 9. A living-room with dining-room adjoining and connected
- ,, 10. Living-room in a house at Malvern
- ,, 11. Dining alcove in the same house

CLOTHES—AND DESIGN — 54 and 55
- Plate 12. The nineteenth century and after, 1801 to 1934
- ,, 13. Cabs, crinolines, overalls and aeroplanes; also men in 1859 and as they may be in 1959
- ,, 14. Street scene, 1862
- ,, 15. Towards functionalism in dress: 1886 to 1934

DESIGN IN THE KITCHEN — 62 and 63
- Plate 16. A German example of sitting-room-kitchen planning
- ,, 17. The latest electric cooker
- ,, 18. Gas stove and hot water set; also anthracite or coke stove
- ,, 19. A kitchen in a country house
- ,, 20. A working kitchen of the minimum flat
- ,, 21. A kitchen cabinet and a pressure cooker
- ,, 22. Combinations of stoves, cupboards and refrigerators
- ,, 23. Roasters and cutlery racks

THE DESIGN OF ILLUMINATION — 78 and 79
- Plate 24. Three types of wall fittings
- ,, 25. A well-lit room seen from outside the house
- ,, 26. Spaciousness emphasized by well-distributed light
- ,, 27. Three types of independent fittings

DESIGN IN MODERN LIFE

DESIGN IN PUBLIC BUILDINGS *Between pages* 86 and 87
- Plate 28. The New King's Road School, Fulham, and the Margaret McMillan Open-Air School, Bradford
- ,, 29. New Liverpool Orphanage buildings
- ,, 30. Astley's Amphitheatre, and the Orpheum Theatre, Golder's Green
- ,, 31. Proposed plan for a sanatorium, and a model of the King's Fund Miniature Hospital

THE DESIGN OF THE STREET 102 and 103
- Plate 32. New and old shopping centres
- ,, 33. A street in Pompeii and a modern street
- ,, 34. A contrast in direction posts
- ,, 35. Lamp-posts and bollards as they might be

DESIGN IN THE COUNTRYSIDE AND THE TOWN 110 and 111
- Plate 36. Planning in the grand manner: an aerial view of Bath, and Wren's rejected plan for London
- ,, 37. Aerial views of industrial quarters and garden cities
- ,, 38. Aerial view of the Hampstead Garden Suburb
- ,, 39. The English countryside and the invading town

ACKNOWLEDGMENTS

For permission to reproduce the progress and period charts on Furniture, Costume, Transport and Utensil Design we are indebted to Mr. Raymond McGrath, B.Arch., A.R.I.B.A., who prepared the latter chart specially for this book. From the editors of *The Architect and Building News*, *The Architect's Journal*, *The Architectural Review*, *Design for To-day* and *The Listener* the most generous assistance has been received in the task of collecting illustrations for the different sections, and they have loaned both photographs and blocks for some of the plates. We are also in the debt of the British Broadcasting Corporation and the Design and Industries Association for their invaluable help in obtaining photographs and granting permission for their reproduction.

The costume illustrations for 1801, 1833 and 1887 on Plate 12, and the 1859 gentlemen's fashions on Plate 13, are from the Victoria and Albert Museum. The modern girl on Plate 12 is reproduced by permission of Matita, Ltd., and Plate 14 is from Mr. Alan Bott's book, *Our Fathers*, and is reproduced by courtesy of the publishers, William Heinemann, Ltd.

The illustrations of tramway stations and signposts on Plates 33 and 34 are reproduced by permission of the London Underground Railways.

J. G.

INTRODUCTION

Most of the chapters in this book have been based on a series of broadcast discussions on "Design in Modern Life" between experts in the various subjects dealt with and a questioner who took the part of what may perhaps be described as the listener's friend. This series of talks was opened by a discussion between Mr. Geoffrey Boumphrey, Mr. Edward Halliday and the writer. In that discussion, Mr. Boumphrey speaking as an engineer, Mr. Halliday as an artist, and the writer as a minor critic of architecture and industrial design, tried to reduce to simple terms the problems that were involved by a general application of design to the conduct and environment of contemporary life. Afterwards, as the talks were broadcast every week by the B.B.C., either Boumphrey, Halliday or the writer acted the part of "listener's friend" with the various authorities who dealt with different sections of design. A small sixpenny pamphlet edited by Mr. Noel Carrington was published when the talks began, and the considerable sale it enjoyed, as well as the interest roused by the talks themselves, suggested that a permanent, illustrated record of this concentrated survey of the disabilities and possibilities of design in modern life would be acceptable to the public. It was clearly impossible for the talks to be reproduced exactly as they were broadcast; they were discussions in conversational form and rather tiresome to read. They have been rewritten, amplified and subjected to complete revision for the purpose of this book. An entirely fresh chapter dealing with the design of houses and tenements has been specially written for this book by Mr. E. Maxwell Fry. The series of talks was broadcast during 1933 under the title "Design in Modern Life." The introductory chapter that follows is founded on a broadcast talk by the writer, the first of an earlier series on "Design in Industry."

JOHN GLOAG

CHAPTER I

WHO KNOWS WHAT THE PUBLIC WANTS?

by

JOHN GLOAG

CHAPTER I

WHO KNOWS WHAT THE PUBLIC WANTS?

THE way to read this book is, to quote *Alice in Wonderland*, "to begin at the beginning and read till you come to the end and then stop." Then think. If you are still satisfied with your surroundings, not your personal surroundings which depend upon your own taste, but the surroundings which you cannot control, and if you are not anxious to make some changes, then you have only yourself to thank if the same dull, ugly and inconvenient ideas are constantly thrust upon you by the people who are providing goods and services in this country, and who think they know what the public wants. Please do not imagine that this book is a piece of highbrow carping at what exists. It has not been written with a sneer at everything English; it has not been written with any conviction that things are always done better abroad than in England. It is written by people who know their own subjects thoroughly, and who know how well we can plan and design in England when we are allowed to. But it is the general public which really controls this matter; that unknown, enormously powerful, strange, anonymous thing, the public, which can make or break ideas, which is coaxed and cajoled by every species of propaganda, and of which no one can say, "*I* know what the public wants." And it is at this point that a great secret may be revealed: the public doesn't know itself what it wants.

This does not mean that we escape the uneasy and irritating feeling that we are constrained to want merely what we can get. Isn't there sometimes at the back of everyone's mind when they go shopping the suspicion that somewhere or other the very thing they wanted must exist, and that if only there had been the time to look for it they would have found it, and not be "making do" with something which isn't really right? Perhaps the only exhaustively critical shopping that is done to-day is by women when they buy their clothes.

There are more women to-day with a sense of style about their clothes, and who obviously choose them very critically, than there has ever been since it was possible for the outer shell of a humble toiler to be almost identical with that of a duchess. But is the same critical care, the same sense of decorative fitness that is given to clothes, ever given to anything else? For instance, don't you, reader, sometimes feel that you ought to be able to see and to buy far simpler and better things for the price you are prepared to pay than the things you actually can see in shops? Don't you think that innumerable articles, whether they happen to be jugs, plates, chairs, tables, wardrobes, radio sets or gramophone cabinets, are just fatuously dull and stupid in their shape, and unbearably fidgety by reason of the type of ornament that is sprinkled over them?

It is the business of this introduction to ask personal domestic questions. Just glance round the room you may be sitting in, or think of any room you know. How many things in it are sensibly designed? Let us begin with sense, not that impalpable thing known as taste. Let us take the fender as the first example. Does it stand above the floor on unnecessary feet, so that it presents you with a dirt-collecting space underneath, or does it sit flat on the same level as the hearth? Is it difficult to clean because it has lots of pierced ornament on the front of it, and a strange growth of lumps and bobbles which are pretending to be ornamental but which are really manufacturers of housework? Are there knobs and points and things which are stuck on, and which in time fall off or get knocked off? Hold an inquest on the fender, and then when you have brought in the verdict of sense or nonsense, be your own coroner about the mantelpiece and the articles thereon. The fireplace probably accommodates a gas fire or a coal fire or an electric radiator. Is the mantelpiece a really good frame for those things or is it covered with flowers and fruit, bits of machine-pressed composition ornament stuck on in ignoble imitation of some antique prototype, ready to catch any dust that is going? If it is an old carved mantelpiece, something that hails from the eighteenth century, then it may be beautiful; but we have got to remember that when it was made, it was probably for a house that had the finest labour-saving machine in the world—a large staff of competent servants. Old elaborate things which may be beautiful, and new elaborate things which are often infernally ugly, are just a nuisance in the small rooms of the modern house.

On the mantelshelf what do you find? A clock? And what sort of a clock is it? Has it been designed primarily as a clock or has it become merely an excuse to introduce a whole lot of ornamentation into your room, the sort of ornamentation which might appeal to a monumental mason, angels and columns in black marble and heaven knows what, in the most frightful congestion, looking, on the whole, rather unsatisfactory? Can the small rooms of to-day really stand a lot of pattern and a lot of ornament? We all have possessions which we put into our houses for sentimental reasons, which are the worst reasons in the world. But if we bought only things which were satisfactory for their job, decorated in a way which made us appreciate the pleasantness of their shape, then our rooms would be much more comfortable and restful.

Is your furniture by any chance pretending to be Chippendale, or Sheraton, or Jacobean, and if so, why? George V, not George III nor Charles I, is on the throne. This is the twentieth century, although everybody who wants to sell you things to put in your house seems to be in a Mad Hatter sort of conspiracy to try and make you forget it. Not only inside your house does that conspiracy manifest itself. Go into the street and hold an inquest there.

Why are you, or perhaps your neighbours, living in an imitation Tudor house with stained wooden slats shoved on to the front of it to make it look like what is called a half-timbered house? Those slats have nothing to do with the construction of the house. They are just applied as ornaments. The house does not look like a real half-timbered house and it never can. It has been built in quite a different way from a real Tudor house, and it has been built by people totally different from the people who lived in the sixteenth century. Again, why has the local petrol pump got a mock Tudor canopy shoved over the top of it? Why is the local tea-shop called "Ye Old Worlde Café"? Why is a local curio shop called "Ye old Gifte Shoppe"? Why do we live in this sort of half-baked pageant, always hiding our ideas in the clothes of another age? People in the eighteenth century never pretended that they were living in the seventeenth or sixteenth century. They were proud of the accomplishments of their time. They designed things for their own day, but we do not let people design things for our own time. Are we ashamed of our period and of all its accomplishments, its machinery, and all the exciting things it does for us? We hide our light under some imitation antique. If you want proof of that, look at gramophone and radio cabinets. One would think that with entirely new things like gramophones and radio we should have given the job of housing their apparatus to a designer instead of to a dealer in disguises. These things were new inventions that started life quite free from any hampering ideas about non-mechanical things that had gone before them. They were absolutely free to be handled in a simple and pleasant way. They were not like railway trains and motor-cars, which have to this day the ghosts of horses running in front of them.

When railways began, passengers were just packed into trucks and left in the rain and the smoke and the grit to get on as best they could during the journey. But wealthy people used to have their horse-drawn coaches hoisted up on to a flat truck, and lashed down safely, so that a vehicle that was really designed for the road was perched like a piece of luggage on a vehicle that was designed for the

WHO KNOWS WHAT THE PUBLIC WANTS?

railway.[1] When covered carriages were made at last, the memory of the coach tied on the railway truck was a fearful nuisance. It prevented people from thinking that a railway coach was an entirely new problem, needing an entirely fresh design which would make the coach really fit for its purpose. Instead everybody remembered the horse coach riding on top of the truck, and in consequence the early railway coaches were made to look almost exactly like two or three horse-drawn coaches sandwiched together. To this day the tradition of the individual coach still lingers on in the individual compartments of most railway carriages, for the saloon carriage is still comparatively rare, and is usually used as a restaurant car.

The ghost of the horse running in front of a motor-car is always visible. The motor-car started life as a horseless carriage. It was always designed as though it were going to be pulled by something. The result was that the early cars were never adapted to the new form of locomotion that they really represented. They were incredibly clumsy, and the drivers always looked as if they should have had reins in their hands instead of the steering wheel. The modern car, with its nice, sleek, streamlined body, has got over the early clumsiness of its ancestor. But it is still being pulled. When the problem of fast and comfortable road travel is tackled purely from the point of view of getting the utmost fitness for function, then a very different-looking car makes its appearance—a car with the engine at the back and a curved, humped body that sweeps over from back to front, cutting down wind resistance to a minimum. One of these cars of the future was shown at a recent motor show, and there are now a few on the streets of London. They have laid the ghost of the horse for ever. Motor-'buses have had the same history. The early motor-'buses had the same form as the horse-'buses. Even the driver was

[1] Mr. Jorrocks arrived by rail at Handley Cross on "an open platform with a broken britzka, followed by a curious-looking nondescript one-horse vehicle" which contained the Jorrocks family.

perched up high above the street, so that the people on the top deck could talk to him, as they used to chat to the old horse-'bus-drivers. (But they didn't talk to him incidentally—it was too risky.)

Those are two examples of what might be called ancestor-worship in design. It isn't difficult to understand why some existing things should have inspired the makers of something that was entirely new. It is only natural to resist new ideas. They disturb the settled routine of laziness. So, naturally, everybody tries to fit new ideas into the existing scheme of things. The ideas, like the unfortunate people who were accommodated on the bed of Procrustes, are lopped or stretched to suit some arbitrary tradition.

It was very easy for manufacturers in the early days of machinery to overlook the possibilities of the new tool that was put into their hands. They only thought of the machine as an accelerator, as a multiplier. They were obsessed by the idea of quantity and they ignored technique. The machine, regarded primarily as a multiplier, was devoted to the imitation of things that had formerly been made by hand, and was never given an opportunity of doing its splendid best. When it was necessary for things to be originated with the aid of machinery it was found that manufacturers left to themselves with their machines could only mix up different sorts of imitations. Machine production never came under the control of designers. We have never planned any partnership between designers and manufacturers.

Not long ago Great Britain went off the gold standard. Everybody heard about it. It came suddenly and it affected our pockets. Just about a hundred years ago we went off what may be called *the design standard*, but nobody heard of it. It came gradually and it only affected our eyes.

Industrial production has been carried on in this country for over a century with very little understanding of the importance of design or even what design is. Respect for design in industry pays as a national policy. France has tested that over a period of two hundred and fifty years. Why do English manufacturers go to France to get ideas for their patterns? Why is Paris the centre of fashion? It is because in the days of Louis XIV, Colbert had the foresight to arrange really practical partnerships between art and industry in the making of all sorts of things, glass, pottery, lace, textiles and so forth. Mr. R. H. Wilenski in his brilliant little book, *A Miniature History of European Art*, has described in two sentences the motives and results of Colbert's policy:

"His aim was to acquire for the French the reputation of the finest artist-craftsmen in Europe because he knew that such a reputation would be a great cash asset to the state. He succeeded; the reputation and the revenue persist to this day."

It is hardly a compliment to our own designers, and we have a lot of talent in this country, that British manufacturers should ignore them and get their ideas from Paris because an overworked French statesman had the wit to make a plan for design in industry two and a half centuries ago.

Manufacturers and designers ought to be partners, but for years manufacturers found it so much easier to imitate old things, and designers for a long time would not realize that the machine was a super-tool and treated the whole of industrial production with contempt. For example, in the middle of the nineteenth century William Morris decided that there was nothing to be done with industry, and that the extension of mechanical production would mean the eventual extinction of craftsmanship. He started a handicraft revival and organized the production of hand-made things which were beautiful, well made and expensive. Unfortunately this only gave manufacturers another opportunity for imitation, and within a few years of the beginning of the Morris handicraft revival factories were turning out intentionally rough things of wood and metal, the latter covered with imitation hammer marks. These things were sold under the intriguing label of "hand-made."

The manufacturers took the line of least resistance; like the designers and artists. It was easy to imitate old furniture and antique designs of any kind, and the imitations became conscientious, and, in one specialized branch of the furniture industry, profoundly dishonest, for the promising market that awaited the faked antique attracted the unscrupulous. For over thirty years hundreds of highly skilled men, capable of original work, have been forced to imitate the work of seventeenth- and eighteenth-century craftsmen. Factories have been organized to produce "Jacobean" furniture, and great furniture designers like Chippendale and Hepplewhite and Sheraton must have turned in their graves at the things that have been perpetrated in their names. How can original design take root in modern industry when it is withered by this old-world cult that has made thousands of people prefer styles instead of sense?

That original design is possible in modern industrial production was proved vividly and conclusively by the exhibition of industrial art in relation to the home, which was held at Dorland Hall in London during the summer of 1933. The holding of such an exhibition followed the recommendations of a committee which was appointed by the Board of Trade under the chairmanship of Lord Gorell. That exhibition gave some members of the public an opportunity of seeing good modern work. It is finally the responsibility of the public to seek out such work. Much of it exists.

Many manufacturers are collaborating with designers, and businesses that have entered into such partnerships have not lost money by their courage. But now comes another difficulty. The manufacturer does not deal directly with the public: his work is bought by the retailer, and the retail buyer is in command of the whole situation. He is more nervous of experiments than the manufacturer. He wants something safe to sell.

WHO KNOWS WHAT THE PUBLIC WANTS?

When you go shopping, everything you buy is chosen for you beforehand always by two, and sometimes by three, people. If you are lucky the first of these people who decide what you will be able to buy when you go into a shop is a designer. He is an artist, and his job is to choose, out of the various equally good ways of making an object, the way that will give the best appearance. He also invents the decoration, selects the colour, and decides the finish of the job. A designer will not allow the pleasant colour and marking of wood to be darkened and stained and spoiled by so-called antique finishes. You will not find him letting dirty colours or shouting colours upset patterns. You will find that in decoration he always knows where to begin and when to stop. If you are unlucky, your shopping will be without the benefit of a designer's watchful imagination. Then you will have only two people choosing your things over your head. Those two people are the manufacturer and the retail buyer.

We can now summarize the chief difficulties in the way of improving industrial design. First obstacle: the manufacturer, who has to be persuaded that partnership with a designer will pay. Second obstacle: the habit many designers have unfortunately acquired of thinking that all business men are fools or sharks or both. Third obstacle: when the first and second have been overcome, the retail buyer, who thinks he knows what the public wants and is nervous of trying experiments. Fourth obstacle: people who shop with their eyes shut, people who buy things uncritically, who allow themselves to be sold things that are unfit for their purpose or that are flimsy imitations of something antique or something incongruous.

Are we going to perpetuate this old-world cult? We are experts at copying the things that Englishmen once designed. But how long will the world be interested in "Olde English" labels? How much have we already lost in the markets of the world by failing to make industrial design stand on its own feet instead of in dead men's shoes? Are we to be respected by the world only for the technical excellence of our workmanship, for our engineering, our locomotives and shipbuilding, and in decorative art and industrial design only for our ability to ape the things our Tudor, Stuart and Georgian forerunners originated? The answer depends on you, reader, as a member of the general public. The answer to every question in this introduction, to nearly every question concerning design raised in the chapters that follow, depends upon you. England, we were told in that wave of war-weariness which masqueraded as optimism after the armistice, was to be made a land fit for heroes to live in. We were again to become the workshop of the world. Instead we have become the slum of Europe.

CHAPTER II

THE DESIGN OF DWELLINGS

by

E. MAXWELL FRY

B.ARCH., A.R.I.B.A.

CHAPTER II

THE DESIGN OF DWELLINGS

We are at the moment of re-enquiry into the design of all dwellings. This must be obvious to any that have eyes to see the world about them. But to recognize change is not enough for intelligent men. That it suffices so many is the reason why movements of great potential usefulness are slowed down and bogged long enough to prevent them from ever serving mankind to their full capacity. Especially is this true of building, where each experiment constitutes an immovable witness for every fool to copy and plagiarize.

The changes in house design now taking place are of the nature of a crystallizing solution at its moment of crystallization; not to be explained without reference to the composition of the solution and to the time taken in arriving at its sudden maturity. Modern architecture took a hundred years to crystallize out—exactly as long as it has taken society itself to shake free from the gross materialism of the last century. And modern architecture and sociology rightly go hand in hand.

We must go back to start the story—beyond the nineteenth century, in search of elementary building. In order to follow this story let me get down to the elements of building and say at once that SHELTER is a primary need, the satisfaction of which, in whatsoever form, is a fundamental pleasure, a constantly renewed miracle. Everything in excess of this is a manifestation of rising intelligence. But even in small units a house has to be planned for bare needs, for comfort, for fine living. And it must be built with the materials most easily available and to the limits of their structural economy. These are practical considerations which, when combined together to give some dignity to the act of living, become Architecture, a background for civilized living. Running over these again, we may condense to Function and Structure, and take Architecture as the indefinable element communicated to both.

Now in the age we are leaving behind us, all these elements of domestic building had got hopelessly fogged, overlaid with association, historical reminiscence and a sort of sentimental story telling. Painting and sculpture had done the same. All the Arts had sickened from this last century of progress. Why?

It is a longish story, but descend again to the elements of building to take this time a concrete example—what is loosely called the half-timbered cottage of an English smallholder of the fifteenth century.

Here is the statement of the programme of this small building:

THE PEOPLE
Very small holders, illiterate farmers, local in every thought and action, thinking without complication of the problems of their existence.

WHAT THEY WANTED
Shelter; warmth, a sleeping-place for man and beast. A storeroom for food and produce.

MATERIALS
English Oak, being plentiful, but hard to work and often twisted. Clay and Earth.
Twigs and Straw for roof, and for Clay reinforcement.
Glass in very small sizes, and expensive. A little Iron, painfully wrought. Brick, not always local.

THE STRUCTURE, i.e. What they did with their materials
Rough framing with selected straight and curved timbers, making the structure rather like an upturned ship. The spaces filled in with

PLATE 1
THE DESIGN OF DWELLINGS

The elementary type of cottage: a simple timber-framed structure, with wattle and daub filling and thatch roof. Picturesque, cramped, mean-windowed. The house of the landsman—built of materials lying and growing within the local radius, put together simply for simple needs.

PLATE 2
THE DESIGN OF DWELLINGS

The house of the social townsman—its materials standardized and made apart: its construction a fitting counterpart of classic taste. (*Photo: copyright F. R. Yerbury.*)

PLATE 3
THE DESIGN OF DWELLINGS

The townsman's house—organized for the community into formal street architecture, each sharing the benefits of a unified scheme. (*Photo: copyright F. R. Yerbury.*)

Another form of sharing that came with the Industrial Revolution—

PLATE 4
THE DESIGN OF DWELLINGS

—and in course of time helped to produce the loss of mental balance that was responsible for this.

PLATE 5
THE DESIGN OF DWELLINGS

Meanwhile city congestion with the aid of steel, concrete and machinery gave birth to the multiple-flat dwelling for the rich, such as this (Flats at 29, Abercorn Place, London, N.W. Architect: Digby Solomon, F.R.I.B.A.—

PLATE 6
THE DESIGN OF DWELLINGS

—and, after further labour, for the poor also. (*Füchsenfeldhof, Vienna.*)
(*Reproduced by courtesy of the Architect's Journal.*)

This form of house is now an anachronism and a fairy tale for tired business men. Wilfully so.

PLATE 7
THE DESIGN OF DWELLINGS

While the modern materials that make flats make the houses that suit our need to-day.

PLATE 8
THE DESIGN OF DWELLINGS

The completely freed interior of a modern house; light, airy, warmed, comfortable and cultivated. (*Reproduced by permission of Miës Van der Rohe.*)

wattle and daub, i.e. clay and earth for the mass, twigs and straw for the reinforcement; small window spaces, because glass is dear, or was dear, and ceilings low. Thatched roof, clay floor.

This is simple building, direct and lowly. It is the plan upon which field shelters are built in the same districts to-day. It is the simplest solution of an elementary problem, done with the utmost economy of effort and material in the light of its day. Note that the resulting cottage was cramped, subject to rising damp, but unartificial, easy to appreciate, harmony rather than contrast.

English domestic architecture was never for long as simple as this, for in Elizabethan times the country was beginning its steady march towards wealth and world conquest. But for a long time its materials remained unchanged, while the technique of timber construction, growing amazingly clever and complicated, degenerated into tricks, and finally fell away before the onrush of a new civilization which came from Italy into the North. And this new breath of life, coming at a time of very material expansion and access of wealth, and coming also at a time of social disintegration into fixed classes, brought to domestic architecture complications of living which entirely transformed its root programme. This change is important enough to warrant a restatement of the problem which, in the small cottage, appeared so simple. Here it is.

THE PEOPLE, now to be called Society

    Classes divided into aristocracy, wealthy and lower classes. Safety of life established. Domesticity formularized. Manners in process of formularizing. The printed word spreading throughout. Religion broadening, and over all this, the dissemination of the idea of fine living upon a background of classic and Italian culture.

ARCHITECTURE
   This is a new interposition. There was for this age only one architecture, and that, classic. It gave the complete background to its culture. The cultured lived for it. They were prepared to find the materials with which to make it possible. They created the architect as the arbiter of taste and master over the works.

MATERIALS, i.e. What they found
   Stone and the new standardized Brick.
   Tiles, and later, Slate.
   Native hardwoods, and then imported soft woods from Russia.
   Lead for roofs and gutters.
   Glass in larger standard sizes.
   Wrought Iron.

Out of these they produced perfectly harmonious building. There was nothing that the needs of the time required that could not be met by a combination of classic forms and contemporary structure, neither doing violence to the other.

The important thing to notice is that modern society, as we know it, was established after the Reformation, and that so long as materials remained unchanged there was no break in the tradition of classical building established at that time. This tradition of building, which we know as Queen Anne or Georgian architecture, was, besides being cultivated, eminently practical. Standardized in its units and very largely in its materials, and capable of being multiplied in straight formation so as to form the matrix of the large-scale planning. The contemporary principles of town building, indeed, were mentally a good deal in advance of our own to-day. Not only Wren's plan for London, never executed, but most of the town developments which were carried out from the date of the Fire of London to the end of the first quarter of the nineteenth century are among the most remarkable achievements of civilized man.

But the whole thing broke into fragments. We were once in command of the situation. We were cultivated. We could build towns on a grand scale. There came the nineteenth century and all this scale and all this culture disappeared entirely. For one hundred years, architecturally speaking, we became either as the brutes or like fools that could not reason enough to use the materials at hand as logically as the fifteenth-century peasant did. It is from this confusion of mind and matter that we are now, and only now, beginning to emerge. Just think what it meant, this nineteenth century. It produced on the one hand an enormous under-educated population of semi-slaves. It gave birth to a powerful class of shop-keepers and merchants. It reduced what was left of a cultured aristocracy to cultural impotence. It played havoc with agriculture. It severed our contact with the earth. It fostered great but unplanned aggregations of factories and dwellings. All this it brought about while, on the other hand, its ceaseless invention was providing us with iron, with steel, with machinery, with the means of draining and surfacing great masses of building. It gave us, in fact, the materials and the means to make an industrial life possible. It gave everything—but the capacity to analyse our wants or our resources. And unable thus to see what we were about, the poets and the artists, looking in horror at the ravages of industry, turned inward and created their own mental picture of what they would like life to be. For close upon one hundred years, domestic architecture was divided into the dream pictures of the rich—"an architecture of escape"—and the meanest forms of low-class housing. No one worried very much about the poor man's house, but they worried a great deal about everything else: worried themselves into every form of revival, into every pretence to hide the reality of industrialism—the steady thrust of a new structural system of steel and glass, and, finally, of concrete.

In vain their worry; for society, grown from eight to forty millions, demanded for the satisfaction of its wants buildings and structures of a scale which only these new materials could provide. The Forth Bridge, Crystal Palace, St. Pancras Station roof, the Eiffel Tower, and behind these, factories, railways, power stations and a myriad of great works, pushed remorselessly, turning architects into a race of peasants, scholars or dreamers, and engineers into architects.

The Garden City movement was the last despairing effort to escape from the new industrial life: from the control of the machine. It is doomed. We are at the moment of complete reorganization, in control of the machine, in control of a new way of life.

Here comes the last restatement. What are we about to-day? What are we going to do with the dwelling? Let us put this down in order.

SOCIETY

Recognized at last as being one, of people having a right to live healthily and happily. As a result of one hundred years of battle, we are nearing the communal way. It is imposed as a duty on the State to see that everyone is housed; that everyone has the means to live, to be educated, to be looked after in sickness and in old age. We have definitely decided that our life must be in the main urban, backed by an industrial system, and served, but not controlled, by machines. We have given up the idea of unlimited personal service. We recognize the right of personal freedom. We are, when we can master all this, more than a stage beyond the admirable culture of the Georgians.

(ARCHITECTURE)

Merely put in here, to be transferred below. It is no longer an interposition.

MATERIALS AND STRUCTURE

All materials now come through the industrial system. All, or nearly all, are standardized. Planned and made for their purposes in domestic building. At last we recognize Steel, Concrete, Glass and Machinery as fit for all building. We are prepared to use them to their greatest economy in the service of the structure, which once more and after so long an interval will arise from the functional use of fitting materials.

Let us examine at closer range. In the immediate past, houses were built on a different plan from flats and hotels. House design relied too much on personal service. It made a point of using old-fashioned materials and a highly subdivided plan of small rooms and small windows, even while it felt powerless to withstand the revolution that was taking place in the bathroom and the kitchen. An American once said that we should one day give the house away with the plumbing. The point is, of course, that the bathroom and the kitchen now come to us directly from industry. They are the products of machinery, and further than that, they are to a growing extent actual machines. Machines are, things into which an enormous amount of thought has gone before being made in great numbers.

To use them efficiently, their position must be planned exactly. The kitchen must share their exact nature and become a machine-room, a room of kindly, helpful machines, designed to simplify and make work enjoyable. All work in the house emanates from here, but because by using machinery and by exact thinking we can compress, the kitchen becomes smaller; it is run by one person. From the kitchen and through the house we see mentally a series of paths leading to the various functions of house running. We study these carefully; we reduce their length, increase their efficiency so that one or two persons can do what six did in the past. We plan our new dwelling exactly. Fifty years ago the working end of a house was to the living end in the proportion of 1 to 4; we can reduce this proportion to one of 1 to 8, and add the rest to living.

In like manner we compress the bathroom, in which we have

controlled plumbing out of sight, because we are sure of it, not because we do not like the look of it. Bedrooms, from the vast caverns to which our ancestors used to creep to shut themselves further into four-posters, have shrunk to rooms in which we sleep, warm and ventilated, our belongings fitted scientifically into accessible cupboards built into the house.

These are changes. We have redivided the house into a compact working space; a space given over to the act of sleeping, planned to be no bigger than it need be; and the rest, spacious, subdivisible at will, is there to live in spaciously, served by controlled heat, made of beautiful but easily cleaned materials, thrown open to the air in fine weather—a flexible, airy dwelling.

The programme of house living has changed, the structure is changing, even the family unit is breaking up and rejoining in new formation, as witness the idea of hotel and hostel living and the emergence of the one-room flat.

How to deal with the aggregation of dwelling units in Urban formation is the besetting problem of this age. It might be considered to be a town-planner's problem, but in point of fact the planner of urbanism is an architect, or that mixture of constructor, designer and sociologist that is to serve society under whatsoever title.

The control of the disposal of dwelling units, whether they be in formation upon the ground, in series of horizontally and vertically linked floors built up from the ground, or in isolated towers with but an anchorage in earth, is our job. The present dispersal wastes all our powers and defeats us at every turn. Our uncontrolled urbanism is a nightmare, to disperse which will require of us the firmest belief in our power of analysing our needs, and having defined them, of bending every contemporary resource to the final building up of a new way of life.

CHAPTER III

THE LIVING-ROOM AND FURNITURE DESIGN

by

GORDON RUSSELL

PLATE 9
THE LIVING-ROOM AND FURNITURE DESIGN

A living-room with dining-room adjoining and connected. By Raymond McGrath, B.Arch., A.R.I.B.A. The curtains enable the rooms to be screened on occasion. (*Reproduced by courtesy of Country Life, Ltd.*)

PLATE 10
THE LIVING-ROOM AND FURNITURE DESIGN

HARMONY OF CUBE AND CIRCLE

This living-room is in a house recently built at Malvern. Movable furniture is reduced to a minimum. (Architects for the house itself and for the interior: R. D. Russell and Marian Pepler. Furniture and hangings by Gordon Russell, Ltd.)

PLATE 11
THE LIVING-ROOM AND FURNITURE DESIGN

Dining alcove of the room shown on the opposite page.

CHAPTER III

THE LIVING-ROOM AND FURNITURE DESIGN

IN most houses the living-room is that room in which we spend our leisure time, entertain our guests, read, write, play the piano or listen to the gramophone or wireless. Its functions, and their variety, really suggest that it is a return of the mediaeval hall, which was the communal room used for all activities of the house in the fifteenth century. From the sixteenth to the nineteenth century, the number of rooms in houses increased and the hall decreased in size. The dining-room, sitting-room, parlour, drawing-room, morning-room, study, library, ballroom and others arrived. Lack of space, economy and difficulties of domestic help have now reversed the whole position, and it is quite common to find a sitting- or living-room to-day with a dining space in an alcove.

Although separate rooms for separate functions were desirable and convenient, certain developments of modern life have changed our conditions of living enormously; namely, the spread of sport and games, and of course the motor-car. *Much less time is spent in the house than formerly.* Our sitting-room is spreading outwards to the loggia, sun-porch, garden, and even to the golf-links and open road. Our cars have indeed become our sitting-rooms for quite a long period each year.

Take another example: the large plate-glass windows which wind down into the walls in some modern German rooms emphasize this connection between the living-room and the open air. They frame up a piece of country as if it were a picture on the wall, thus bringing it into the room when the window is closed; open it and the room goes out to meet the countryside. Social customs are changing, too: entertaining is difficult and expensive. Dining out is being succeeded by "taking coffee out." It is, therefore, easier to merge the dining-room in the living-room, and the extra space enables a bridge party or a dance to be arranged at short notice. You must remember that the small house and cottage have more rooms than they used to have. It is the large house which has shrunk. Also the size of families is much smaller.

This concentration of functions in a room must affect the whole character of furniture design. It alters the designer's approach to that problem.

There is a demand for less and smaller furniture. It is one way of getting more space at a time when rooms are shrinking. Then the use of the dual-purpose piece of furniture will be seen: the table-bookcase, table-stool, even bed-settee which converts the living-room into a bedroom for the unexpected guest, thus taking us a full cycle back to mediaeval times.

A further way of saving space is to get built-in furniture, and this is being done to a much larger extent than formerly. Its disadvantage in the rented house is obvious, but even there it is gaining ground and saving space. Many houses and flats are now fitted as a matter of course with kitchen cabinets and refrigerators, hanging cupboards for clothes and so on. Besides saving space usually taken up by the back and sometimes the sides, fitted furniture is very easy to keep clean—it has no spaces for dirt to lodge in. A still further way of saving space is to plan each piece of furniture so as to fit into it exactly what is required. This may sound obvious, but in very many cases it has only recently been done. Adjustable shelves in bookcases come to mind; in bedrooms most people are now familiar with the fitted wardrobe; in the kitchen, the kitchen cabinet; and in the living-room, the writing-desk, cocktail cabinet, perhaps canteen for silver. Low furniture makes a room appear large. In some cases unit furniture may prove more useful than fitted furniture. Unit furniture is made to a unit of size so as to be interchangeable. The unit bookcase is familiar to most people, and the principle ought to be extended. Someone thought of making a writing-desk which, on its base, would take up exactly the same space as two unit bookcases. It is a very convenient arrangement to be able to write within easy reach of one's books of reference.

Such standardized furniture need not produce any boring or monotonous effects. The infinite variety of arrangement to meet varying shapes of room makes this unlikely. A vast number of variations could be built up with only six patterns: cupboards, with and without doors, writing-desk, chest of drawers, bookcase in two sizes, one double as large as the other, deeper for larger books. And again, why not combine fitted and unit furniture? Book-shelves are cheap to fix and one can think of many pleasant arrangements. Interesting pictures, pieces of sculpture, pottery, curtains, will give intimacy and individuality to such a room. Mass production will supply our everyday needs, but it must also raise the standard of living and education and give us more leisure. It may also lead to a much greater demand for fine individual things made by hand.

Unfortunately, many people at present look upon their living-room as a sort of museum—full of bits and pieces of no earthly use except to fill space and only adding to the labour of cleaning and sometimes making the room a trap for the unwary. A living-room should be as spacious as possible. In some houses it may prove a good thing to take down the partition between dining- and living-room. But mere size is not enough: it should not be fussy. It should provide a restful, un-self-conscious background. It should be well lit, sunny, the walls covered with a light and cheerful, untiring material. Many light colours in distemper or paper are fast and washable, so are curtain materials. Electric light will be carefully arranged to give adequate lighting for reading in several places in the room. This will not be as easy as it sounds, for many electric fittings are not designed to light a room. If the room is centrally heated it will probably have an electric or gas fire, which is useful as a focal point in the room rather than for heat. But have a clean modern design, not one of "ye olde electryc logge" types. It is curious that some people like to have things disguised to look like something else. I suppose they then appreciate the enormous ingenuity of the designer. At the Paris Exhibition of 1900, a piano was shown carved to look like a tree. The child in us applauded such a crystallizing of the fairy story. But it is merely childish. A room does need a point of main interest and a fire does supply this. It may be just tradition. There will be no more furniture in the living-room than is necessary for comfort. The increasing complexity of life leads us to demand simplicity in our surroundings. It is so much more restful than the overloaded rooms of fifty years ago; it is also much easier to clean a room of this kind. The bric-à-brac and curios will no longer be littered over the tops of all furniture. These tops will be useful for the things we use every day, or think we do—papers, pipes, cigarettes, books, things which make a room look lived in. The chairs will be really "easy": this is a characteristic of our times. We no longer sit upright on hard seats, and some of us have even given our living-rooms the undistinguished title of lounge! Just as the complexity of life to-day leads to simplicity in surroundings, so the pace at which life is lived makes comfort at home essential. Few of us fail to demand, and some of us achieve, physical comfort, but the lack of mental comfort is not perhaps so obvious to everyone.

There are competent designers in Great Britain to-day, there are enlightened manufacturers, there are shops which try to set a standard in the things they sell, there are members of the public who care about these things. But they are in the minority. There is, however, evidence that a very large public exists which is vaguely uneasy. It feels it is missing something. It is. It is missing beauty in the things it uses every day.

Ugliness does affect everyone. It is so common to-day as to be taken for granted by most people. To start the planning of better towns we may as well look round our own living-rooms. As with towns, so with living-rooms, there are very few which would not be improved by throwing things out of them. There is no better exercise than to go round one's rooms at intervals and try to look at everything in the room separately as if one had not seen it before. William Morris said: "Do not have anything in your house which you do not either know to be useful or believe to be beautiful." This is a good standard to go on. It is wrong to imagine that experts or art critics only are qualified for this form of scrutiny.

Appreciation of good design can be taught, and there is no better way of learning than by looking at things which informed opinion has approved over several generations; things which may be seen in plenty in our great museums. To be able to form an opinion of any value on the work of to-day, it is essential that we should study the best work of yesterday. But we must regard this work as a standard of design, not as models to copy. And also remember that not all old work is good.

The taste of the consumer should be educated so that there will be a larger demand for well-designed things, and the machine will then be used as it should be—to turn out cheaply and by tens of thousands examples within the reach of the slenderest purse. After all, the much maligned machine is already doing this where it is controlled by men who understand its peculiar limitations and scope. It is turning out modern things, not half-baked and indigestible copies of traditional things. There is, naturally, a certain amount of distaste for modern things. Every new idea has to fight a battle for acceptance. But most *young* people do not dislike their own period. In many ways, it is a great age. Possibly any prejudice there may be is really against "modernistic" design rather than modern. Just as designers without either culture or originality have borrowed the more superficial decoration of various periods of the past, so to-day they content themselves with taking the superficial motifs of good contemporary design and thinking they have "mastered the tricks of this modern stuff," they produce the horrible "modernistic" lights, fabrics and furniture which are seen everywhere to-day.

It may be taken as a very healthy sign that such spurious modern design is disliked by so many people, even though the same people go in for what are in fact equally deplorable imitations of the past. Good design will please you after seeing it continually for years. Meretricious work soon palls.

There is furniture which is good, furniture which is fashionable, and sometimes furniture which is good and fashionable. Let us start criticizing from the point of view of whether a certain table is fitly designed for its job. Does it stand level? Is the top smooth and without cracks? Is it a convenient height? If it is an occasional table, is it light enough to be moved? If a dining-table, do the rails bark our shins or the frieze damage our knees? Is it strong enough to sit on? for few pieces of furniture are used solely for the purpose for which they were designed. All this is sheer common sense, and if we had more of it and less talk of periods and styles we should get much better furniture. It is time we made an end of this tyranny of imposed taste, or rather snobbery which passes as taste. Everyone can decide to think for himself. He will soon see that things are as badly designed as they are, not because the makers think they look better like that, but because they haven't thought at all. Just laziness—the same laziness which induces the customer to buy the article which is put before him, even when he knows it to be bad, rather than search for something better. The public is becoming steadily more critical and will certainly get what it demands. It is annoying to be considered a crank because one criticizes the design of an article, especially as design should start from the very beginning; and if it is good the completed article will be more efficient—design is not something added at the end.

The general standard of design in housing is lamentably low, but certainly improving. This is perhaps responsible to quite a large extent for much of the poor furniture, curtain materials, carpets

THE LIVING-ROOM AND FURNITURE DESIGN

and other things. Jerry-built, jerry-designed houses naturally contain things that match.

The builder and furniture designer supply a demand: they can't afford to do otherwise. The public must *demand* good stuff. *Laissez-faire* won't get us anywhere. Shopkeepers pride themselves on knowing what the public wants: but does the public itself know what it wants? You can hardly expect a shopkeeper to stock things to fit every individual taste; but what we ought to expect is that each piece is good of its kind and for its price.

The aim of designers for mass production should be to produce furniture which is functionally perfect and pleasant to look at in as small a range as will meet most needs. It would be possible to spend quite a lot on the designs in the first place. The manufacturer should get the very best designer he could; the cost of design would add very little to each piece as so many would be made. The problem of the specially designed piece of furniture is the same as the made-to-measure suit—mass production offers us the suit or the furniture ready to wear.

Anyone who is going to furnish a living-room should write down their requirements. This clarifies one's ideas. Make a plan of the room, showing which way the windows face. Be sure to sit in the room: to walk through a room gives such a different impression. Don't buy anything until you know what you want, and then buy the absolute minimum, but buy it good. As in cooking, it costs so little more to buy the best materials, but avoid waste like the plague. Apart from the satisfaction of the room's functions, there must be decorative relief. There is a definite need of pictures, if only to humanize a room where the furniture achieves a severe beauty by being made primarily for use rather than for show. The pictures themselves must be of a very high or very low standard if they are not to fade into the walls as it were and remain unnoticed. But they must not overcrowd the walls.

If you collect pictures, why not put most of them, like books, in a library? That is, either keep one room in the house as a gallery and plaster it with pictures, or store them away in cupboards and take them out and display them periodically. Some focal point in the room may be reserved for a picture—over the fireplace is a good position—and one should be taken out of the store and set there and left until it has been read, then it could be replaced by another. In this way the pictures would really be looked at and not merely taken for granted as part of the traditional furnishing of the room. In the same way it may prove useful to have one or two beautifully designed and hand-made pieces of furniture—antiques if you like—in a room which is otherwise furnished with things made by the machine. The two techniques of making are quite different, and both, in a properly balanced world, should be good.

No civilized country can think lightly of scrapping handwork. A great deal in the work in cabinet-making, for instance, is still *better* done—more efficiently done—by the hand than by the machine. It must of necessity be expensive, and few people appreciate the niceties of cabinet-making sufficiently to judge between first-rate and poor stuff. What is very important is that we should not look upon machine-made furniture as an inferior substitute for, or imitation of, handwork. Both have their own characteristics, and real advance will be made when we realize this fully, as we are beginning to do. All designers worth their salt want the public to be much more critical of their efforts. To say that the English are not good at designing things is wrong. At one time we built houses and made furniture which have hardly ever been surpassed. We are essentially a home-loving people. There is a great stirring of the public conscience about slum clearance, bad housing, poverty, disease, spoiling the countryside. Ugliness is always the enemy—bad design is one form of ugliness. It is up to public opinion to fight ugliness in all its forms.

CHAPTER IV

CLOTHES—AND DESIGN

by

JAMES LAVER

CHAPTER IV

CLOTHES—AND DESIGN

DESIGN does not apply to the creation of clothes in the way it applies to modern decoration or modern architecture. Although clothes are the most intimate of personal possessions, the character of a man's clothes is wholly impersonal. Most men simply wear the uniform of their profession and are quite content to do so; and the uniform of the Stock Exchange or the Law Courts is as unmistakable as that of the Post Office or the Police Court. In fact it is easier to recognize a stockbroker or a lawyer than a postman or a policeman, for you often know them even when they are off duty.

Outside the labelled professions, there are what used to be called the leisured classes. Being—to use an old-fashioned phrase—a man about town consists in little more than wearing the clothes a man about town should wear. Then there are certain indefinite types of uniform: the conscious eccentricity of the budding artist; the carefully careless exterior affected by all manner of people engaged in creative work of some kind. They aim at comfort and usually achieve untidiness; nevertheless theirs is a professional uniform. A sort of squalid functionalism, if a term that seems so architectural be permissible.

Functionalism has little to do with the established and stable and respectable professional uniforms. Tall hats are not the ideal headgear for travelling to the City in; and stiff cuffs are still worn by clerks in spite of the fact that it is difficult to write in them. One could multiply such instances. Plumbers wear lounge suits, although lounge suits are quite unsuitable for their life—I almost said for any life, although the lounge suit is undoubtedly a step in the right direction, after the frock coat, and the morning coat, which still survives, and until a few years ago was quite common.

Consider modern clothes (men's clothes, that is) merely from the point of view of suitability. Think of collars, stiff or soft. Think of hard hats leaving a red dent on a man's forehead. Think of waistcoats contracting his chest and exposing his kidneys. And trousers! What can you think of a garment that has to be hitched up every time a man wants to sit down. Men are suffering prisoners in their formal clothes to-day. That does not imply that nobody is well dressed. It is possible to be well dressed while being functionally slightly ridiculous. The doctrine of fitness for purpose and pleasantness in use breaks down badly when we analyse sartorial psychology. It cannot be denied that it is always pleasant to see a man well turned out, with clothes that fit and are made of good material. It is possible to be well dressed in the clothes of almost any epoch. But no one, I think, could look at the clothes worn by the Romans or the Elizabethans, or the men of the eighteenth century, with their knee-breeches and three-cornered hats, and pretend that modern clothes are beautiful. One has only to look at some of our public monuments to see that marble trousers are painfully absurd. Cloth trousers are sufficiently ludicrous. I wear them myself because Convention is stronger than Comfort, or rather because comfort consists of other things beside bodily comfort. In my own case the comfort of not wearing a collar would be counterbalanced by the discomfort of being stared at. Most men, including myself, do not feel strongly enough about these things to expose themselves to ridicule. We are all governed by the tyranny of Good Form. Not that it's entirely a bad thing. It's one of the ways of helping a lot of people to live together in comparative comfort. It involves our National psychology. In spite of social changes, socialist movements, standardization, popular journalism and what not, the dominant ideal in England to-day is still the ideal of the English Gentleman.

For more than a century and a quarter the English gentleman has imposed his code on England and his clothes on the whole civilized world, and being an English gentleman he is suspicious of change.

He is contemptuously suspicious of "movements" such as Dress Reform. He has a horror of cranks, for although we have been a nation of pioneers we have never been able to establish a reliable border-line between legitimate pioneering and crankiness. The English gentleman, whose clothes are a pattern to the world, is too conventional. If he alters the number of buttons on his coat sleeve—or the opening in his waistcoat—he considers he has done something rather daring. His leadership is based on reticence. When men followed gracious and decorative and exciting fashions a century and a quarter ago, it was not the Englishman who led them. Until just before the French Revolution it was the French aristocrat who imposed his fashions on the civilized world. Till then, Paris dictated not only women's fashions but men's as well. Then, with the collapse of the Old Regime, the growing taste for country life instead of the Parisian salon, and the general admiration for England as a land of liberty and prosperity, English male fashions overcame the French, and have never lost the position they gained then.

Those English fashions were devised for a country life, particularly for riding. That was the idea of the cut-away tail coat, which remained as a day coat until about 1860, and as an evening coat until the present day. That was the idea of the hard hat—to protect the head when thrown from a horse. Those things have a functional origin.

Trousers are more difficult to explain. They were worn by the *sans-culottes* of Paris; they were worn by the sailors in Nelson's fleet, but they didn't make their way into fashionable attire until later. Then they were a kind of dashing eccentricity; the sort of thing young bloods like Byron wore. There is a story of the Duke of Wellington being refused admission to a fashionable club because he was wearing trousers.

Although these early nineteenth-century clothes were country and riding clothes, they were not so inappropriate, even in towns, as we are apt to think. There was an enormous amount of riding during that period. But even if there had not been it would not have made any difference. All the newly rich classes wanted to be thought English gentlemen, and country gentlemen at that. So country clothes became stereotyped. Most curious of all, the hard high hat, the top hat, became absolutely universal. It was a kind of mark of respectability. Cricketers wore it; it was worn by both crews in the Boat Race; and when we still want to look very respectable—as at weddings or funerals—we wear it still. Even the bowler, when it first came in, was regarded as rather rakish and fast, and it is only in our own time that the soft felt hat has swept the world. Before, it was only worn by farm labourers, socialists or poets.

About the middle of the last century a sudden drabness came over men's clothes. One of the reasons was, of course, Good Form, which in one way is a kind of upper middle class conspiracy. Your aristocrat is naturally ostentatious. He likes to be conspicuous. It was a great victory for the middle classes, just coming into power, when they managed to make ostentation seem bad form, even a trifle vulgar. Then, when differences of rank have disappeared, social conventions tend, for a time at least, to become more rigid. The only way you can show you are a gentleman is by dressing with a certain quiet distinction. Snobbery is the essence of the matter.

Another cause of drabness was the nineteenth-century smoke cloud. There has always been plenty of dirt in the world, but, for the first time, in the nineteenth century the very air was dirty over large tracts of England. Coal was being burned in increasing quantities; factories poured upwards a sooty canopy over industrial areas. Hence dark clothes, detachable collars and cuffs that could be changed frequently, and a new enthusiasm for baths. The nineteenth century had to make more fuss about cleanliness than the eighteenth just because its world was growing so dirty. Central heating and electric power may dispel the industrial smoke cloud in our own century, and the Central Electricity Board may be the unconscious sponsor of brighter fashions for men. Already there is a movement in that direction: brighter pullovers, coloured blazers, white flannels, light-coloured suits in summer. But it is doubtful whether men will ever be as gorgeous again as they were up to the end of the eighteenth century. They have ceased to compete.

The end of the eighteenth century saw the first movement towards the emancipation of women. Before that time, women—decent women, that is—were kept very much at home. As soon as they began to go about the streets in large numbers, men retired from the contest. They threw the burden of dressing for effect on to their women, who were quite willing to take it up.

A woman is never more convinced that she is expressing her personality than when she is following the latest fashion. Fashion is so important to a woman. Men's clothes are hardly affected by fashion at all, or very slowly. Most men have probably been wearing the same evening clothes for the last ten years, and even if there are not the right number of buttons on the sleeve, only a tailor or a gigolo would notice it. Men have accepted uniforms. Women have demanded fun.

Fun enriched by an element of snobbery; and an element of seduction—for clothes are certainly intended to make women more attractive. Then there is something that, for want of a better name, I can only call the Spirit of the Age.

Women dress to attract men. That is the seduction element. It is apparent even when they do it unconsciously; most of all, perhaps, when they do it unconsciously. Regard any personable young woman of a marriageable age; she may say she doesn't want a husband, but she certainly behaves as if she did.

Indifference is just what a woman must at all costs avoid. Take for example the German nudists—people who spend their holidays in the country and do not wear any clothes at all. Everybody thought (as the Nazis now think) that the German nudist camps would be one long orgy, but they were not so at all. On the contrary, I believe there was much less flirting and lovemaking in them than in the ordinary English seaside boarding-house. Once nudity ceased to be a novelty it very soon became a bore, or, at least, a matter of indifference. And indifference must be avoided. Clothes lend an element of mystery, and they also serve to emphasize one part of the body at the expense of the rest. The part varies in every age. In 1800 it was the bosom; in 1860 the shoulders and the waist; in 1870 the hips; in 1900 the ankles; in the nineteen-twenties the legs. And it varies because, just as the whole body is a bore if it is always visible, so any one part of it becomes boring after a time. The great leg-epoch is a fresh memory. It's over now, and nobody seems sorry. Nothing definite has yet taken its place, unless it is the bare back in evening frocks.

There is a flavour of functionalism in the design of women's clothes. A man who designs a speed-boat designs it for speed; the man who designs a woman's dress designs it to be attractive. A speed-boat has not only to be fast, it has to keep afloat, and a dress has not only to be attractive, but to keep the wearer warm—although if a woman has to choose between keeping warm and being in the fashion, she'll tell you that she's hardy by nature and never feels the cold.

Women often dress to cut out other women in the competition for the attention of men. Men may not notice the details, but they are very conscious of the general effect. Other women copy the details of a successful toilette, and still more women copy them. And so the new, striking original dress is copied and copied until it loses all its appeal. It becomes a bore through too much repetition. Then, when

PLATE 12
CLOTHES—AND DESIGN
THE NINETEENTH CENTURY AND AFTER

Empire costume, 1801.

Evening dress, 1833.

Bustle, 1887.

The modern girl.

PLATE 13
CLOTHES—AND DESIGN

Narrow cabs and wide crinolines.
(*Reproduced by courtesy of the proprietors of "Punch."*)

Functional dress for the modern airwoman.

MEN'S DRESS, 1859 AND 1959 (?)
Showing the type derived from English riding costume of the end of the eighteenth century; and Dress-reform costume.

PLATE 14
CLOTHES—AND DESIGN

Street scene, 1862. "The crinoline was at its widest in 1860, when 'buses and railway carriages were twice as narrow as they are at present." (*Reproduced from "Our Fathers" by permission of the publishers, William Heinemann, Ltd.*)

PLATE 15

CLOTHES—AND DESIGN

TOWARDS
FUNCTIONALISM
IN DRESS

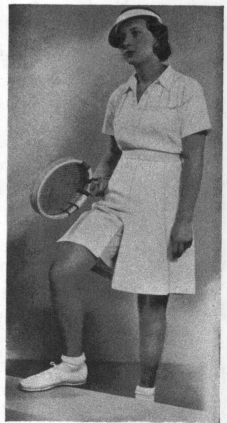

1886 to 1934

(Reproduced by courtesy of Lillywhite's, Ltd.)

it's been adopted by the definitely unfashionable classes, no fashionable woman dares to wear it. She has to find something new.

With modern mass production, fashions filter down very rapidly, and that is why they change so quickly. The little bowler hat of a couple of summers ago appeared quite suddenly, but it was too simple. A fortnight later the shops were flooded with them, so that the fashion was killed almost before it was born. There *are* a few fashions which last—but it is their attractiveness that appeals before their utility. Small, close-fitting hats and berets for instance. There was, of course, a sporting note in the beret, and you could wear it at any time except on the most formal occasions.

Most women's fashions start in Paris, but many fashions start which nobody takes up. That is because of that intangible something that I can only call the Spirit of the Age. To explain that, let us forget about the fashions of the last few years. They are too near to us to be seen in perspective. Think of earlier fashions. Isn't there a curious suitability about the stiff brocade dresses of the late seventeenth century? About the powdered hair and huge panniers of the mid-eighteenth? When the French Revolution swept all that away you got simple, classical dresses rather like nightgowns. Then by reaction there grew up an excessive prudishness in dress, so that the early Victorian lady felt almost naked if she didn't wear seven petticoats, and frozen if two of them weren't made of flannel. The naughtiness of the 'nineties gave us the *frou-frou* skirt and lace-edged underclothes, carefully displayed. In the short skirts of a few years ago you can see the triumph of the post-war flapper determined to have a good time. As soon as you can see any Age in perspective, everything hangs together, but nobody can prophesy what the present Age is going to look like to posterity.

In 1928, or thereabouts, Paris designed long skirts—and nobody would wear them. So Paris went very cunningly to work. Strips of material were added to the sides; a train was devised which left the skirt short in front. Then the long transparent over-skirt was invented, which Paris finally ventured to make opaque. Also the Spirit of the Age was turning in their favour. The wild, extravagant 'twenties were coming to an end. The party was over. The world was suddenly a serious and rather anxious place. I think that in a few years' time it will be quite obvious that the disappearance of women's legs coincided with the beginning of a new era—a kind of new Victorian period. I was prophesying it three years ago when skirts were at their shortest. These changes come, and the most the dress-designer can do is to anticipate them and try to lead the way.

In spite of modern conditions, cars and 'buses, I wouldn't put it past women to return to all manner of cumbersome horrors. The crinoline was at its widest in 1860 when 'buses and railway carriages were twice as narrow as they are at present. There were plenty of 'buses, and even motor-cars, in 1912, but that didn't prevent women from wearing hats as big as cartwheels, kept in place by enormous —and dangerous—hat-pins.

Sport may well save us from the worst extravagance of fashion— sport and the new interest in machinery. Women might ride in a brougham even if their crinolines stuck out of the windows; they could hardly fumble for the clutch through seven petticoats; though even now their evening dresses get tangled in the gears. But then evening dresses are not designed for day-time functions, although they have nearly always been simply a more seductive version of day dresses.

Evening dresses for both men and women may disappear altogether. If Communism became universal that would probably happen, because then the weight of snobbery is thrown on the other side of the scale. But if that doesn't happen, I think it more likely that some kind of evening dress—party dress, if you like—will become universal, at least for women. If you go into the cheapest dance-hall, where there's not a boiled shirt to be seen, you will find that most of the girls are wearing some kind of evening dress. And the dresses they wear, owing to mass production, get better and better, and apparently cheaper and cheaper. But I think we shall see the end of formal dress in the daytime. The morning coat will become the uniform for reception clerks in hotels and undertakers. Otherwise it will vanish. It has almost gone already. The lounge suit will replace it, for as a suit it is certainly more functional than the dress of the last generation. And it has one very important advantage over dungarees and golfing jackets and pullovers and sweaters: its pockets. Civilization has given us so many little things to carry: lighters and cigarette-cases, and fountain pens, and dyspepsia tablets. The lounge suit is the nearest approach to functional dress that we are likely to get, barring revolutions, and remembering that every sport already has its proper uniform—or will have. There is a growing tendency to play tennis in shorts, and golfers may stop wearing heavy tweeds, which were suitable enough for the Scottish climate and have gone all over the world because golf is a Scottish game. So much for functionalism with men.

I repeat that women's clothes *are* functional, that their function is to make their wearers attractive, and they certainly succeed. Some women's fashions may seem ugly in retrospect, but never at the time. The man who thinks the fashions of ten years ago more attractive than the fashions of to-day is simply getting old.

Children's clothes can and should be functional in the simple sense of the word, and they generally are nowadays. We no longer think it necessary for little boys to be as *respectable* as their fathers, nor that little girls should be dressed in their mother's finery. Gym suits for girls, by all means, and flannel shirts for boys, and for babies, rompers, or better still, nothing at all. It will be time enough for them to dress like their elders when the boys are beginning to earn their living and the girls are thinking of getting married.

But however much we may talk of function design in clothes, even as an ideal, there is something in the matter which for ever escapes us; dangers which can never be foreseen but which, when seen in retrospect, seem part of the essential character of an epoch. The apparently trifling nature of changes in fashion is what makes them of much importance to the historian and should make them of such interest to those who are endeavouring to shape the new world. For fashion is like a weathercock which shows which way the wind is blowing before more solemn and serious arts are aware that the wind has changed.

CHAPTER V

DESIGN IN THE KITCHEN

by

ELIZABETH DENBY

CHAPTER V

DESIGN IN THE KITCHEN

THE design of the kitchen is scientifically studied to-day when large institutions and stores are built, while in domestic architecture many are beginning to consider it as the centre and not merely as an adjunct of the house. This view of the kitchen as a vital domestic workshop is bound in time to affect the planning of all types of houses: but it is unfortunately true that expensive and presumably authoritative books on architecture may not contain a single illustration or even mention of a kitchen; exhibition flats are sometimes shown without one at all; or if a kitchen is included, its meagre equipment is a stove, a deal table and a chair. This does not suggest much general constructive planning or creative interest in the subject yet, and it is because of that absence of interest—especially official architectural interest—that in many instances the basic planning of an inconvenient kitchen is more at fault than the disposition of the individual pieces of equipment.

The kitchen should be the initial responsibility of the architect, since it is essential that its building, design and arrangement should be considered in relation to the rest of the house before the house is built. After all, the kitchen is a working unit which must be equipped to meet the needs of a normal family in the most economical way. It must be planned to solve the problem of getting work done as quickly, as well and with as little effort as possible; the equipment thus needs to be a carefully thought out part of the room and not a series of afterthoughts inserted at the discretion (or indiscretion) of the housewife.

Here is a list, a formidable one, of work that must be done in a kitchen: firstly, of course, the storage, preparation, cooking and serving of food; secondly, the housing of the pots and pans; thirdly, the supply of hot water for household use (and in many houses for laundry work); fourthly, the management and cleaning of the whole house, which must be carried out from here; while finally, the kitchen is generally a dining- and sitting-room for the maids—or, in the normal small house, for the whole family.

Work must therefore be minimized by considering and simplifying the sequence of operations in preparing food and other household tasks. This secures economy of movement: If the relation of equipment is properly planned, a cook can be in the middle of her kitchen and yet have all she wants at arm's length merely by turning round. Some labour-saver in America once counted the steps a cook took in making a cake—they came to 281, which by merely rearranging the kitchen were reduced to 45!

The majority of kitchens which are also used as sitting-rooms would be improved by being divided into a cooking and a living area. People rightly expect comfortable and pleasant surroundings for their leisure time, and if the less agreeable jobs are not to spread about the room, the stove, sink and larder must be carefully grouped; this would incidentally do away with much wasteful effort.

ASPECT

The kitchen should, if possible, get the morning sun; the idea that it must face north was to keep it cool owing to the heat of the cooking ranges. For every degree of heat they put into an oven, they put six into the room and the cook's face, and an incalculable number into the cook's temper! Improved insulation is altering this, and there is already at least one fuel cooker which wastes so little heat that it would not keep a large kitchen warm enough in cold weather.

LIGHTING AND VENTILATION

A kitchen cannot be too light, and windows can be helped by the use of light-reflecting materials and pale colours on the walls. A glass panel below the actual opening of a casement window, for instance, allows the sill to be used and increases the light. Windows should be easy to open and to clean, and a fine gauze screen should be fixed in summer to keep out flies. The top of the window should reach nearly to the ceiling in order to carry off hot stale air and steam. Proper use can then be made of the ceiling as a reflector of light, while the elimination of that unnecessary strip of shadow along the cornice makes the whole room seem lighter. As to the type of window, in spite of some obvious disadvantages I personally prefer casements, as they are easier to open and to regulate. They also admit more air and light, but one of the drawbacks of metal windows is that the standard size is so short that a choice has to be made between seeing out and having a dead air space at ceiling level; the taller sizes are considerably more expensive.

Sash windows of course have their supporters, and they certainly do permit of window-boxes—a town substitute for a garden—that gives pleasure inside and outside the room. (Sash cords, which are a source of trouble in an old house, last very much longer if they are steeped in tallow before they are used.)

After windows, independent ventilation should be considered. Smells from the kitchen must not penetrate to the rest of the house. Air bricks just below the ceiling, a properly designed oven, and most important of all a hood over the stove with an extractor into the flue, are the most effective ventilating methods.

Lighting of equipment is more important than is generally realized, though the grouping must depend to some extent on whether the room is to be used as a kitchen only or also as a living-room. If the former, the sink, refrigerator and stove could be in a line below the window, with the top of the refrigerator available as a table. In a sitting-room-kitchen, the sink could be on the side wall next to the window, with the stove beside it. They would each be well lighted from the side, and this arrangement would leave the window free.

Electricity is the cleanest and coolest artificial light and points should always be placed in proper relation to stove and sink, so that the cook is not in her own light. Fitments are now designed which are a great improvement on the bulb dangling from a piece of flex, and which take much less cleaning. An example of efficiency from a foreign kitchen is a light on a movable flex, so that one light serves instead of several, being moved along the flex to where it is required.[1]

STOVES

When choosing a stove, the first consideration should be the kind of cheap fuel available in the district. Anthracite, oil, gas and electricity are alternatives to coal, and their use will help to diminish our smoke problem. Gas and electricity have the further advantage of needing no fuel store; electricity and oil need no outside flue.

A fuel range is automatically included in the fittings of the average house; but gas and electrical apparatus must generally be bought or hired separately, and the stiff fees charged for hiring electrical equipment is often the deciding factor against using this form of power.

The number of stoves on the market is bewildering. Most of them are ugly, with decorations which are survivals from last century and with a pathetic belief in the necessity for legs, whereas more oven space would be available and less dust collected if they rested solidly on the floor. An inexpensive stove which is simple and decorative, which will economize fuel by conserving heat, which will cook well, need little attention and be easy to clean, is still wanted. The only one which combines these advantages is unfortunately very expensive to install, though incredibly cheap to run.

[1] For design in lighting see the next chapter by A. B. Read.

SINKS

The next key piece of kitchen equipment is the sink. Two jobs so often need doing at the same time—putting clothes in steep, washing-up, washing one's hands, soaking vegetables—that a double washing and rinsing sink is invaluable. Taps should be high to prevent breakages and should be of a sensible shape, cleaning being now saved by the use of chromium-plating or hard enamel. Self-washering taps save cost in renewals, while big enough draining boards and an adequate plate and cup rack (the latest pattern has a gully for catching the drips and leading them back to the sink) should help to break down the general hatred of washing-up. A good device for washing-up consists of two basins, one on top of the other, with a strainer between. When the plug is taken out of the top basin, the scraps go into the strainer and the dirty water is caught in the bottom basin.

A shelf for soap, brushes and other utensils and a strainer should form part of the design of every sink. A water softener saves soap and cleaning materials in a hard-water district, but it is still beyond the pocket of the average householder.

As for materials: porcelain is as good as anything at present, though stainless steel and other materials are sometimes used. Breakages are easier in a metal or porcelain sink, but a wooden surround and a rubber mat at the bottom are said to be effective protection.

LARDERS

As dry goods can be kept in the kitchen, and tradesmen's deliveries are increasingly frequent, a larder can be quite a small affair nowadays. Refrigerators will some day make a separate larder for perishable food unnecessary: meanwhile they are costly luxuries.

Larders should of course be on the north wall and be well ventilated, though some protection is necessary against dust blowing in from outside. Shelves should be of a cold material such as slate, marble, tiles, and it ought to be unnecessary to say that no hot pipes should be run near the larder! In a small kitchen the dresser can be ventilated on the side against the outside wall, one cupboard being thus the larder while the other can be used for china or other stores. This saves space and looks pleasant.

Wire racks which keep vegetables fresh and clean are preferable to other methods of storage. Some are now made with a tin base to catch earth and dead leaves, while others are constructed on castors and have a table top, so that they can be wheeled where they are wanted.

PANTRY, STORE-ROOM AND FITTINGS

The store-room and the pantry are two of the rooms which are now being telescoped into the kitchen, by the excellent method of lining the whole of the available wall-space with shelves and cupboards in which everything can be kept clean and orderly. Besides being a real saving in space, it is much easier to be tidy with adequate cupboard room. But if cupboards were not provided when the kitchen was built, well-designed fitments are now on the market comparable with the unit system of living-room furniture: Fully equipped kitchen cabinets, dressers, china cupboards, broom cupboards, ironing boards which fold into the wall, flap tables, and tip-up, built-in seats can all be obtained and fitted into the most convenient position in the kitchen.

Designers would do well to remember that mouldings and ledges harbour dust, that cupboards are better shallow, and that at least half an inch should be left between shelves and wall to make cleaning easy. Tube racks, which do not hold the dust, make very convenient pan rests. Glass drawers in a kitchen cabinet, big glass sweet jars with air-tight stoppers (which cost less than a shilling each) or screw-capped glass canisters (for threepence) enable the condition and quantity of stocks to be visible.

A swing-door is convenient in a kitchen, as one's hands are generally full, but if handles are preferred, levers are easier to manage than knobs and a small ledge of wood fixed on the wall by the door will act as a tray-rest while the door is being opened. A serving hatch or lift between kitchen and dining-room saves much traffic, particularly if space is utilized between the double doors for keeping those things which are in constant use for the table.

A funnel to the outside dustbin from a point just below the sink is very useful, as refuse can be wrapped in newspaper and thrown directly through the aperture into the bin. If a stove is used in which suitable rubbish can be burnt immediately, the refuse pail can be dispensed with altogether, and in any case there are better containers than the open bucket!

Public laundries, the bag wash, and perhaps especially the simpler clothes and linen now in use, are doing away with the need for big washing appliances in the home. The heavy mangle is being superseded by a light and portable wringer that screws on to the edge of sink or bath. Airing cupboards round the hot-water cylinder, drying cabinets, airing racks on a pulley over the sink are all available. Electric and gas irons are nearly foolproof, while ironing boards are designed to slip under table-tops or hinge behind doors or be put into any available and convenient corner.

UTENSILS

Utensils are the tools of the kitchen. They should claim attention after the workshop has been well planned, lit and ventilated and provided with clean and abundant power. Pots and pans should be chosen because they are easy to clean, economical in use and in addition good-looking. The old-fashioned cast-iron pans which could hardly be lifted, even empty, are almost extinct. Stainless steel, enamel, aluminium, fireproof pottery and glass are all easy to clean, while the great advantage of glass and pottery is that food can be both cooked and served in them. Tiers of pans which economize in fuel and space are now finding a formidable competitor in a pot which cooks a whole meal at a time.

Generally speaking, cooking utensils are sensibly, if not always beautifully, designed. They are really fit for their purpose, and if they are sometimes unpleasant to the eye, it is because needless additions in the name of decoration have destroyed the clarity and simplicity of their form.

FLOORS

The finish and decoration of the kitchen have a profound influence upon its practical comfort. The floor may easily wreck an excellent plan if it has a tiring or a slippery surface. Here again there is an abundance of materials. Cork is perhaps best, although comparatively expensive. Whatever material is used, however, it is wise to choose a colour which will not show up the dirt which is always brought into a room. Chalk and sandstone country, for instance, would need very different colours if foot-marks were not to show. If the skirting is painted black, it will not show brush marks, and if the floor is curved up into the wall or stove or cupboard base, it will be much easier to clean.

WALLS

In a well-planned kitchen there would be very little wall-space unoccupied by cupboards or fittings, so that practically all the surface would consist of wood and glass. Paint and general finish should certainly be in light and cheerful colours—for instance pale blue, clear green and light grey instead of the "serviceable" dark brown too often chosen.

There are also many easily washable wall materials which can be used; and even if tiles are not generally used because of their rather hospital-like appearance, they should *always* be used as a splashboard for the sink and be fitted round the cooking stove. American cloth

DESIGN IN THE KITCHEN

is now specially made for pasting on like wall-paper, and this is quite good if the proper adhesive is used.

Curtains can be used for decorative effect, but should be easily washable—such as gingham or oiled silk, rubber and american cloth which can be scrubbed or wiped over with a cloth.

EQUIPMENT

The equipment of the kitchen can be completed by a good strong table with drawers beneath, a flap extension and a glass or enamel top to save scrubbing. Extra table space can be provided by flaps hinged on to the walls or by a chest of drawers on castors which can be wheeled where needed. A stool that can be converted into a pair of steps for reaching high cupboards, one or two strong chairs, and some easy folding chairs for leisure use are the remaining essentials.

GENERAL

To summarize the influences that should control the initial design of the kitchen, the functions of the different parts must be analysed and the fittings regarded in relation to each other.

The cooker, sink and larder are, as it were, the three focal points. If they can be close together, so much the better, and they must in any case be placed in relation to everything else in the kitchen and in alignment with each other.

At the sink vegetables are prepared, dishes are washed-up, scraps and trimmings thrown away, so the vegetable rack and the refuse pail should be near—and the obvious place is under the sink or just under one of the draining boards. Similarly, the position of equipment in relation to the cooker should be controlled by the cooking operations entailed. In cooking, ingredients are collected, prepared and mixed either in a pan or in a dish, cooked, taken out and dished up. The larder should not be too near the cooker; but everything else can and should be—mixing table and store-cupboard, so that once fish, poultry or meat is on the table, everything else is handy —pots, pans, mixers, mashers, cooker and dish all within a step or two. The food can then be prepared and cooked without any unnecessary movement; all that remains is for it to be dished up and taken to the table; and that should not mean a long journey even in a large house, if the plan of the house has been properly thought out.

If a method of logical analysis were used throughout in the design and planning of every detail—what a kitchen would result! "Saki" once wrote, "She was a good cook as cooks go, and as cooks go, she went!" Would any cook really abandon the perfectly planned kitchen with well-designed equipment?

Nobody knows yet.

DESIGN IN THE KITCHEN

A German example of sitting-room-kitchen planning. The working section can be curtained off from the rest of the room. The french windows lead on to a pleasant large balcony, where children can play while their mother is engaged in her housework.

To the right a close-up of the "kitchen" end shows how cleverly space has been utilized. The draining board folds over the sink when not in use. The alignment of sink, dresser-top and cooker should be noted.

(From *Eine Eingerichtete Kleinstwohnung*, by Franz Schuster, Hoffman, Stuttgart.)

PLATE 17
DESIGN IN THE KITCHEN

The latest electric cooker here illustrated has ample space for roasting, boiling, grilling, and for heating plates. The well-lighted double sink and draining boards with cupboards beneath, the water heater and refrigerator, are other convenient features in this kitchen designed by Raymond McGrath, B.Arch., A.R.I.B.A.
(*Reproduced by courtesy of "The Architectural Review."*)

PLATE 18
DESIGN IN THE KITCHEN

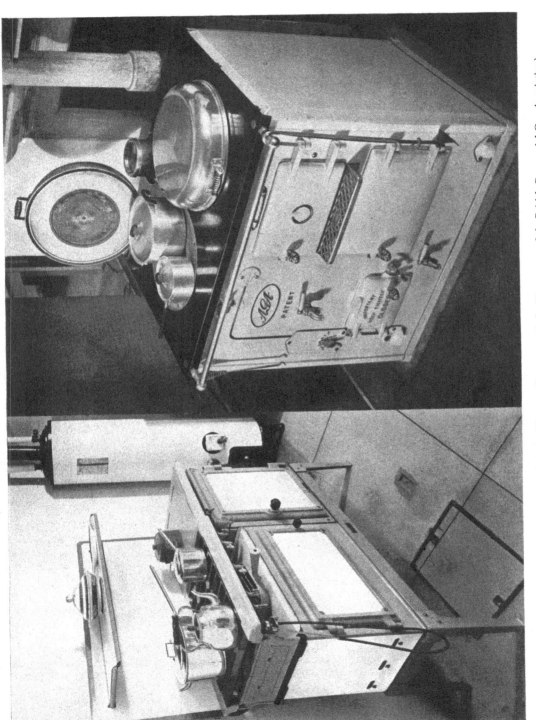

Left: Gas stove and hot water set. Engineers: Thomas Potterton, Ltd. (*By courtesy of the British Commercial Gas Association.*) Right: Anthracite or coke stove. This AGA cooker is so well insulated that the normal consumption of fuel is 22–25 cwt. per annum, and the kitchen is kept at a comfortable temperature. Engineers: Messrs. Bell's Heat Appliances, Ltd.

PLATE 19
DESIGN IN THE KITCHEN

Above: Kitchen in country house. The painted ceiling and tiled walls and floor are extremely easy to keep clean. The window and glass-panelled door give a feeling of light and cheerfulness to the room. Note the glass panel below the actual window opening, which allows the window-sill to be used without lessening the light; the double draining board; the mixer tap; the cupboards below the sink; and the clothes airer. (Designed by W. F. Crittall.)

Opposite: Working kitchen of the minimum flat, designed by Wells Coates. The overall size of this room is 5 ft. by 4 ft. 8 in.; the housewife has one square yard of space to move in, yet she probably has more actual "room" than in many kitchens twice the size. Note the cupboard below the sink; the built-in drawers; the relation between the stove, sink and larder. The equal height of these three ensures economy of movement, all are well lighted, and the top of the refrigerator forms a convenient table. The wall cupboards provide ample storage space, while things in constant use can be taken quickly and easily from the one open shelf. The draining board catches drips from the plate-rack, there is an anti-splash surround to the sink, and the walls are of washable paint. This kitchen was shown at the exhibition at Dorland Hall in 1933. (*Photograph reproduced by permission of Wells Coates.*)

PLATE 20
DESIGN IN THE KITCHEN

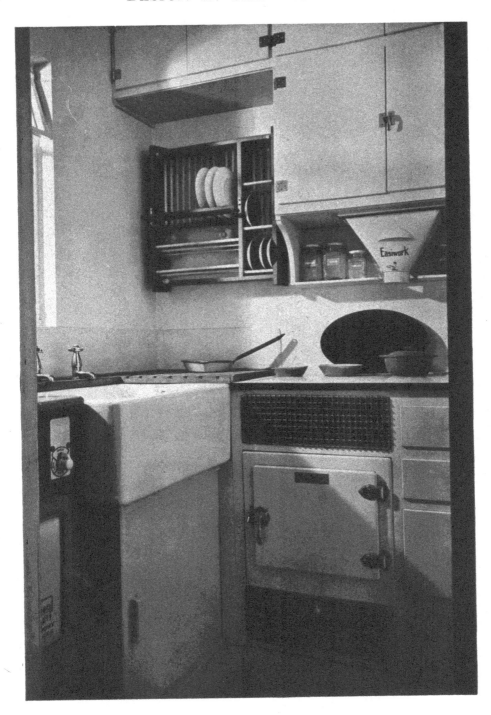

PLATE 21
DESIGN IN THE KITCHEN

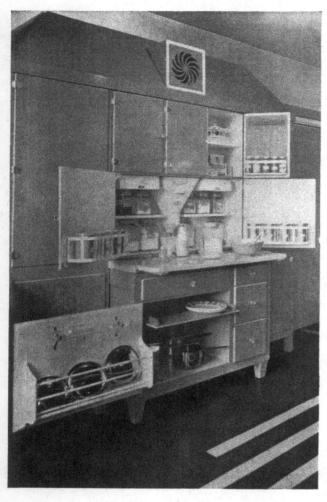

Above: Kitchen cabinet, designed and made by
Easiwork, Limited.

Right: Pressure cooker, automatic cooking by controlled steam. The containers enable a complete meal of various courses to be cooked simultaneously.
(*Reproduced by courtesy of Easiwork, Limited.*)

PLATE 22

DESIGN IN THE KITCHEN

Above: A compact combination of stove, refrigerator, cupboards and drawers. The amount of table space above the larder and again in the dresser—both the same height—should be noted.

Left: An excellent working table has been obtained by combining cupboard and refrigerator into one fitting. The cupboard uses every inch of space to the floor, but beneath the refrigerator creates a real sweeping problem. (*Reproduced by courtesy of "The Architectural Review."*)

PLATE 23
DESIGN IN THE KITCHEN

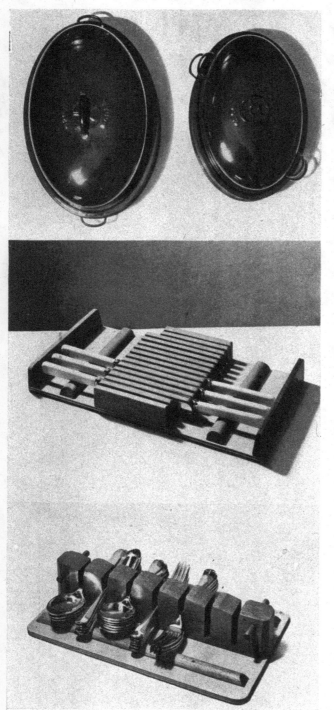

Top: Roasters. (Supplied by Staines Kitchen Equipment Co., Ltd.) Made of sheet steel and finished in hard enamel, these can be used for almost every form of cooking. These utensils embody features which are worth studying. The position of the handles on the right-hand model occupies less space in both height and depth when used in an oven. An outstanding example of real design.

Centre: Knife rack. (Designed and made by Staines Kitchen Equipment Co., Ltd.) For collecting knives, saving handles, etc.

Below: Spoon and fork rack. Pat. "Silver Tidy." Designed and made by Staines Kitchen Equipment Co. Ltd.) Cutlery stored in these portable trays can be easily counted, saving handling and facilitating table-laying.

CHAPTER VI

THE DESIGN OF ILLUMINATION

by

A. B. READ

A.R.C.A.

CHAPTER VI

THE DESIGN OF ILLUMINATION

To keep within the most convenient bounds of probability, the design of artificial illumination should be examined in terms of electricity. We are now at the beginning of the new age of power, the age of clean and abundant power, for the Grid is completed, and it is only a question of time before archaic methods of generating power and providing artificial light disappear. This will mean that our towns and cities and countryside will escape for ever from smuts and soot and fumes; that by day we shall get more and clearer sunlight, and by night more abundant light.

Electricity will be available in most parts of the country at a reasonable cost in time. This cannot happen at once, unfortunately. The Grid was planned to transmit electrical energy in bulk, and to link up the various undertakings already distributing it. At the moment it is, of course, rather exasperating to see pylons in one's garden and yet not have the supply in one's house. But it should be remembered that most of these are high voltage transmission lines, and that they cannot be tapped anywhere and a supply suitable for domestic use obtained. The actual distribution of electricity is an intricate system, and it will be many years yet before every district can possibly be served. One or two very thinly populated areas, such as parts of Scotland, have purposely been left alone for the time being, owing to the small potential consumption of current, which could not possibly justify the enormous cost of running these overhead lines. Judging from comments in the Press, there are still many people who so hate the sight of these towers that they would rather exist in the gloom of candles than avail themselves of the boon of electricity.

Incidentally the expression of violent opinions about these towers has made us forget all about telegraph poles; yet when they were first erected only a small percentage of people could see them as anything but eyesores, and ignored any good they might bring.

We may assume that when there is a universal appreciation of the peculiar gifts of electricity there will be so much of it used that where power lines are run over open country, funds will be forthcoming for them to be taken underground. Meanwhile people still have to learn how to use electric light.

It is amazing the number of lamps with brilliant filaments that are even to-day quite unshielded. Often, too, these bare lamps are hung at eye level, which is most trying to the eyes. Where they are not covered or enclosed with diffusing shades then pearl or opal bulbs should be substituted. Bare filaments should never be visible to the eye. That is an elementary fact that is too often unobserved or neglected.

The designing of artificial light must begin with the house plan. When a house is being built there seems no possible excuse for not planning it, and lighting it, intelligently; but attempt to improve an existing one, and it is much more difficult. Lighting points, like many other important things in the house, always seem to be in the wrong places, and the mind that puts the fuses and the meter in the innermost and dimmest recesses under the stairs, forcing us to lie flat on our stomachs to mend a fuse, can hardly be expected to put lighting points in sensible positions. Altering existing points often means the cutting of walls and a general mess, so that if changes are considered they should be made when new decorations or furnishing is being done. Of course, great changes can be made by using existing points and scrapping the fittings. This is the easiest and possibly the cheapest way to improve an existing system. It is possible now to buy lighting fittings that have been designed and have not just happened. We have to remember that there are millions of homes where there is no money to spare for fittings for lighting as we generally consider it, and where people are fortunate if they are able to afford even enclosures for the lamps themselves. In all such homes what is wanted is a variety of a few very cheap, enclosed fittings, perhaps made of white opal glass, providing good light, simple, yet good in shape, and completely avoiding any pretence at being decorative.

Designers and architects have done a great deal to encourage the standardization of simple lighting fittings and we can now buy British-made fittings which though cheap and efficient are worthy of a place in the smallest or the greatest houses of the country. Perhaps the possibilities of making the bulbs in more pleasant shapes will eventually be explored.

Far too much has been done to make light a decorative thing. There has been too large a choice; in fact, to a lot of people, lighting has become a toy. Big theatres and cinema corporations have handled it to produce effects which can only be described in the terms of film publicity as "spectacular," "stupendous" or that old favourite "super." This has really very little to do with sensible everyday matters of lighting, for in all good design the simplest solution, properly explored, is generally the right one. It is probably better, viewed from every standpoint, to use one of the simple fittings just described, than to worry ourselves by lighting our rooms by many of the cunning and curious methods which are now in vogue. Naturally, where the general standard of living is on a higher level we cannot exclude the right use of lighting in any form, and where means allow us to light a room from more than one source, then agreeable atmosphere and various forms of decorative interest can be achieved.

It should be remembered, too, that walls and ceilings must be coloured to help the lighting of rooms, if we wish to be intelligently economical. Clean, pale colours reflect most light, and it is expensive to supplement light in rooms whose surfaces are dark and dull in tone. It would be cheaper in the long run and infinitely more satisfactory to make the walls and ceilings a lighter colour.

It is difficult to do much to existing homes without incurring fairly heavy expense. It is rather like paying to alter a badly fitting suit of clothes after it has been worn for a year or two. It can be improved, but can hardly be expected to be as good as a well-fitted new one. It is a comparatively easy matter, though, to put in a few more lighting plugs in the skirting, giving us an opportunity of employing floor standards, which throw the light towards the ceiling, giving a diffused light over the room, or to use table lamps, which can serve as local lights and yet at the same time give a really pleasing light generally.

Indirect lighting, with the source of light altogether hidden, practically doubles the consumption of current, and unless properly used it has a rather drowsy appearance. One of the difficulties of indirect lighting is that everything tends to become lit in a monotonous way. One does want a certain contrast of light and shade and it is actually more restful.

Speaking generally, it is better to decentralize the light, or at least to have points provided so that light can be obtained from sources other than one. The reason for this is that quite often the centre of interest of the room may be a fireside or a window, and the sense of comfort and friendliness can be helped by lights here and there where they are needed. By having several points of light, the shapes of the furniture and of the room itself can be made to look more interesting than when illuminated from one quarter only. This means, of course, slightly more expense both in the installation and in the provision of the lighting units themselves, but for most people of moderate means this is the general principle of lighting, which we most of us want, and some of us get.

In a new house, the problem of lighting can be considered with freedom to make the best possible use of the means available. Passing over the purely technical question of where the supply is brought into the house, it should be laid down that the distribution board and fuse boxes should be in an accessible place. Perhaps it is easier to approach the problem from the outside, as it would be an advantage if we and our friends could see through the gate and reach the front door without groping. A light near the door, preferably illuminating the number of the house, would prevent a lot of cursing and stumbling.

In the hall there should be a light sufficient but not too bright, because one does not want the living-rooms of the house to be dim by comparison. There should be lights in hat and coat cupboards that would switch on and off with the opening and shutting of the doors. Our faces could be inspected and the gruelling effects of a hard day seen in the lavatory mirror if the light here was over the mirror and basin, instead of on the ceiling behind our backs. In passages, staircases and landings, where there is never too much room, light should be as snug as possible, tight on the walls or ceilings; neat tubular fittings or (better still) light provided from indirect sources, leaving the walls and ceilings clear of projections.

The sitting-room should have a comfortable, restful, general light, with local lights for reading and sewing. There should be good illumination, but no glare, because we want rest and comfort, with no distracting fussiness to take our minds from the conversation or our book. We should like to feel here that time has stood still for a little while, and that we are right away from the rush and disturbance of our everyday lives.

The source of light should be as inconspicuous as possible. It is better to look across a room than round a pendant. Lighting must be unobtrusive, for every inch of space is valuable, and must be used without waste. Light must inevitably become more and more a part of the form of the rooms themselves. Except in very high rooms the lighting fitting should be as close to the ceiling as possible, preferably tight up to it.

It is only recently that lighting by electricity has been given any real thought. We have got to get into our heads that lighting by electricity is something quite different from all other forms of lighting. Why lighting fittings should hang down, imitating oil lamps and candle fittings, passes my comprehension. Our houses were once filled with these things. Now the trade in imitation wax candles must be a very depressed one, and we are beginning to see that flames that never flicker, and drops that never drip, are just bits of obsolete sentimentality. Lighting fittings of this sort look about as foolish as automatic cigarette machines with cabriole legs. It is by looking at things in this way that we can improve existing homes, and at least replace fittings of illegitimate design by simpler, neater and unobtrusive units. One cannot over-emphasize the vital importance of this general tidying up and sensible arrangement and planning of the houses we live in. There is a very urgent reason, too, an economic one, behind this general simplification. We cannot afford large houses and flats with ceilings so high they can only be reached on the tallest pair of steps. Our rooms are 8 or 9 feet high, and no longer 12 or 15 feet, as they were half a century ago. There must be space and freedom of movement about the house, and this development of good design, which after all is only based on real common sense, simplifies the interiors of our homes, minimizes the difficulties of furnishing and makes them easier to run.

Bedrooms need a quiet and general light in addition to bed-lights and dressing-table lights. Here again, many of us still put up with the absurd counter-weight contraption which never hangs straight, which does not light our faces, but simply throws annoying reflections and images in the mirror. Apparently it has only recently been discovered that it is more comfortable to read in bed with a light at one's side or over the head of the bed than to have a light hang-

PLATE 24
THE DESIGN OF ILLUMINATION

Top: Wall fitting in chromium steel and opaque glass. (Designed by Antoinette Boissevain for the Merchant Adventurers.)

Centre: Wall light in chromium plated brass and opaque glass. (Designed by A. B. Read for Troughton and Young, Ltd.)

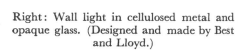

Right: Wall light in cellulosed metal and opaque glass. (Designed and made by Best and Lloyd.)

PLATE 25
THE DESIGN OF ILLUMINATION

The charm of a spacious well-lit room seen from the garden. This house at Poole, designed by Edward Maufe, F.R.I.B.A., is made inviting and cheerful by the use of good lighting and its effect on simple interiors and well-chosen materials.

PLATE 26
THE DESIGN OF ILLUMINATION

Spaciousness emphasized by the use of well-distributed light, sources of which are partly concealed. (Architect: Edward Maufe, F.R.I.B.A.) (*Reproduced by courtesy of the "Architect and Building News."*)

PLATE 27
THE DESIGN OF ILLUMINATION

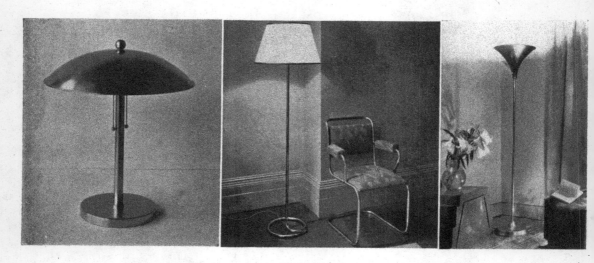

THREE DESIGNS OF ELECTRIC LIGHT FITTINGS

Left: A "direct" fitting by the Merchant Adventurers, which is suitable for local lighting only, for reading or sewing.

Centre: A type of direct fitting by Best and Lloyd with a parchment shade and an open top, giving a soft light all over the room.

Right: A totally indirect floor standard by Troughton and Young, directing light on the ceiling, giving reflected light over the whole of the room. This light is sufficient for all but concentrated work, and is practically shadowless.

ing from the ceiling. In general, the switching arrangements at the head of the bed are antediluvian, and the usual bobbing, dangling switch which is so difficult to catch in the dark should be prohibited.

Then as to the dressing-table lights. Women always ask for a harsh white light; but perhaps *that* is a question for psychologists.

The bathroom is usually the best room in any house because you cannot play about with ridiculous fittings unless you have more money than sense. The lighting required is a good general light from a clean centre waterproof fitting with an additional light for shaving over or at the sides of the mirror.

In the kitchen the operations of cooking and the preparation of meals are intricate; they could be enjoyable if the lights in the kitchen were well placed. A dust-proof, clean centre fitting over the table, with a light over the cooker and another one over the sink and a light in the pantry. These seem obvious things to point out, but in thousands of homes these ordinary matters of lighting have been given no thought whatever. Lights should be arranged so that they shine into kitchen cupboards. In the kitchen and other working rooms the lighting fittings should be business-like in appearance and easily cleaned. The very look of an efficient fitting tends to get the work done better.

In the dining-room, dinner would be a happier affair if the dining-table were well lighted, either from the table itself or from above. In any case, the table should be free from glare, and the rest of the room quietly lighted. Dining-rooms as a whole seem to make one of two mistakes; either they have a brightly lighted table with the room itself too obscure to move about in comfortably, or they are too brightly lighted all over, making the meal a sort of public affair and devoid of all the subtle invitation of a well-laid table. Something between the two is what is wanted. And how rarely one meets it.

Of lighting fittings generally, imitation period fittings are bad enough, but most of those of so-called "modern" design are the worst. Especially those which are the sort of backwash of the 1925 Paris Exhibition, formed of cunningly arranged and vicious-looking jagged pieces of coloured glass, often sprayed a sour-looking yellow, shading rather abruptly to "ripe tomato." There is also the unfunctional functional fitting so mannered and clever, that is supposed to be symbolic of this mechanical age.

Efficiency is the root of the whole question—efficiency, not only in the light given for current consumption, but also in the fitness of the light for its purpose. In other words, the application of that not very common quality: common sense.

CHAPTER VII

DESIGN IN PUBLIC BUILDINGS

by

ROBERT ATKINSON
F.R.I.B.A.

CHAPTER VII

DESIGN IN PUBLIC BUILDINGS

PROBABLY the three types of public buildings that are most susceptible to modern change are schools, theatres and hospitals. They cater for the three things that touch everybody and which are the work very largely of communities: Education, Recreation and Healing.

To begin with schools. Their antecedents must affect their design considerably, and before the actual problem of school building is examined, the origin of schools as they are to-day should be considered. Many of the existing foundations date to the dissolution of the Monasteries. When the Monasteries were broken up there was no system of education left in this country. That is why so many grammar schools and public schools date from the reign of Edward VI. These are what may be called the privileged schools; i.e. the schools where the privileged classes were educated. They date from the first half of the sixteenth century onwards. Not until the nineteenth century was there a change over in education from the privileged class to the masses. The problem of education in London was taken over by the School Board, and certain regulations were laid down about 1867 for the planning and construction of schools.

There has until recently been very little alteration in these regulations, and on the whole the Board of Education has provided first-class guidance for the architectural character of schools. But in recent years the regulations have become largely advisory. They have laid down a certain cost for new schools per head. Only really good designers can work on an arbitrary figure like this and produce admirable designs cheaply. They only are able to produce the best effect from the least expenditure.

Designers of schools, unfortunately, have more limitations imposed on them than cost. The trouble is that architects are often in advance of the regulations imposed upon them. And such regulations weigh very heavily against the primary schools, where cost is everything.

In secondary schools, often called central or county schools, the regulations are much relaxed with regard to buildings that are external to the general teaching, such as manual training, domestic science, and even, where the school is big enough, the provision of a swimming bath or a gymnasium. In the American schools, the opposite number of our primary schools, half the time is given up to the outside subjects, that is to organized play, manual training, housewifery and many things that we don't touch until later in our educational system. Another interesting feature of American schools is that each class-room has its separate cloakroom.

There is also a convention for using rather repellent materials both outside and inside schools. The convention of using dark paint should go. Dirt should be cleaned away—if you can't see dirt you may miss cleaning it. The ideal school should have clean, simple lines. Unfortunately clean, simple lines are regarded as bleak and unfriendly by people living in complicated and muddled homes. There is nothing bleak about simplicity. I recall a school in Spain which had well-proportioned rooms, perfectly plain white walls, not very large windows, pale blue woodwork and nothing else in any room to trouble the plain expanse of wall surface; only a crucifix.

Naturally the whole character of a school building is dominated by the teaching system. Day schools in this country have no social side. They are just a collection of class-rooms and corridors. There is no inducement, indeed no provision, for any pupil to be there a minute before the bell rings for him to be herded into a class-room. This produces a collection of corridors and class-rooms, and a building in the usual arid, treeless setting. There is seldom a place

DESIGN IN PUBLIC BUILDINGS

where pupils and teacher can get any sort of intimate contact out of class hours. It is perhaps because there is little or no social contact that schools look so forbidding. As to "playgrounds," it is a wrong and misleading word. We ought always to think in terms of playing fields. A playground was a luxury when schools were first projected by the School Board, because schools were built in the middle of towns; but there has been a change since the original regulations were laid down which had to make provision for playgrounds. Even roof playgrounds are common enough in congested areas. Most school extension has now shifted (with residential areas) from the centre of towns and cities to their suburbs; where land is less costly, and playing fields either adjoining or very near to the school are possible. Certain modern schools, such as the Handside School at Welwyn Garden City, are arranged so that the class-rooms on one side can be opened on to pleasant gardens, and the children are not necessarily condemned permanently to an asphalt playground. With this change of situation that has taken place, a school building can now become more like its original form, its mediaeval form, which was a monastic building, so that the school is, architecturally, a community, with co-ordinated activities and recreations, and not a type approximating to what a modern block of flats is. It can again become a unit, instead of a series of isolated cells: in short a real community instead of a garage for pupils.

We tend in this machine age to think too much of "parking" people. We park them after they grow up and leave school. In theatres, for instance.

Theatres have certainly lost all their old spirit of fun, the sort of fun the audience made for themselves. The Crowd, the audience, is now herded mechanically into its seats. It isn't given much chance to enjoy "itself." People used to go to the theatre to see and be seen. *There is a real importance in proximity in an audience.* It increases enjoyment. But now there is hardly any showmanship left in theatres, in that sense of showing off the audience, and none at all in the cinema. So many square inches of seating space per person. How can the numbers of the house be increased to the highest limit? That's the chief question asked. Others have to be answered, such as easy access to seats, so that the constant coming and going of the audience in a cinema which is fluid rather than fixed, as in a theatre, doesn't make everyone uncomfortable.

To begin at the beginning with theatres. First there is the queue. Now the queue has considerable psychological value. It whets the appetite of the Crowd—a long queue induces people to enter and indicates success. After all, the herd instinct is uppermost in all amusements—fairs are typical.

It has been suggested that all congestion at entrances should be eliminated, and that something like a passimeter, such as they have in the Underground Railway booking offices, would be better than one pay-box and a turnstile. Also that it would be a good idea to have escalators to take people up from the box-office to the level of the dress circle, the upper circle and the gallery, the escalators being reversed after the show, to get people out of the theatre reasonably quickly. But escalators are wasteful for a theatre. They are designed to cope with fully five thousand people an hour. In a theatre you only have a few hundreds per hour. Also their steps are too big for the L.C.C. regulations for theatres. Normally a pay-box or turnstile can deal with a thousand people per hour.

The queue, then, is a living advertisement for the success of a show.

On this point of external advertisement, is it not rather stupid to design the front of a theatre to look like a Roman temple and then cover it up with posters and moving electric signs, so that the Roman temple disappears behind paper and lamps? It would be possible to incorporate great illuminated posters in designing the façade of a cinema or theatre. Many sites have regulations

PLATE 28
DESIGN IN PUBLIC BUILDINGS

THE NEW KING'S ROAD SCHOOL, FULHAM
One of the typical original London School Board buildings, very well designed for its day and the forerunner of a complete cycle of school buildings.

THE MARGARET McMILLAN OPEN-AIR SCHOOL, BRADFORD
The modern tendency in school buildings is more human and less factory-like. (*Reproduced by courtesy of "The Architect and Building News."*)

PLATE 29
DESIGN IN PUBLIC BUILDINGS

A modern school which is architecturally a community. Architects' model of the new Liverpool Orphanage Buildings, Woolton Road, Liverpool. (Architects: Messrs. Barnish, Silcock and Thearle, F. & A.A.R.I.B.A.)

PLATE 30
DESIGN IN PUBLIC BUILDINGS

ASTLEY'S AMPHITHEATRE
The audience in the eighteenth-century theatre was part of the play.

THE ORPHEUM THEATRE, GOLDER'S GREEN

In the modern theatre, the audience is only there to pay. (*By permission of the British Steelwork Association.*)

PLATE 31
DESIGN IN PUBLIC BUILDINGS

(a) PROPOSED PLAN FOR A SANATORUIM

The octopus plan of hospitals is largely favoured by the medical profession. (Designed by the late Mr. Edwin T. Hall and Stanley Hall—now Stanley Hall and Easton and Robertson.)

(b) MODEL OF KING'S FUND MINIATURE HOSPITAL

A different solution of the same problem. Architects: Messrs. Adams, Holden and Pearson. (*Reproduced by courtesy of Country Life, Ltd.*)

that insist on the buildings erected on them having classic façades, although no regulations are made about bills and signs that may subsequently be put over these academic façades. This is known as the English genius for compromise.

Some of the most recent cinemas incorporate frames for posters in their elevations, and their illuminated notices are part of the design. Notably Messrs. Leatheart & Granger's Sheen Cinema. The L.C.C. regulations affect design, but they are wise and not repressive. On the question of safety, the L.C.C. regulations for theatres and cinemas are altogether excellent. They are lucid and most intelligent. The reason is possibly that the plans of theatres and cinemas that have to pass the L.C.C. are now interpreted by architects instead of by engineers.

But the design of theatres has been affected chiefly by a different point of view—the prices for seats. In the old days there used to be a great variety of prices, and each price had to have its separate pay-box and approach to the auditorium. Now, with non-stop revues and continuous performance cinemas the number of pay-boxes and routes to the auditorium from the street have been reduced; the whole theatre plan simplified. Far less intimacy between the audience and the stage has, however, resulted from these changes.

The old-fashioned theatre had a kind of sympathetic connection between the stage and the audience. Theatres in the old days were small. There was a pit for the cheaper seats, and this pit really was the centre of the theatre. It was sunk, as it were, in the middle, and it was surrounded by several tiers of boxes or galleries for the better-class patrons; each gallery or tier was shallow and held comparatively few people. But the many tiers gave a great number of front seats and the house was always "well-dressed" in consequence. That was when showmanship included the audience. The audience was part of the show, and the house was always very familiar, and the stage was so near that the reaction upon the audience was easily gauged, and the gags were coloured to suit. There are several of these old playhouses still left, almost in their original condition, in Bristol and Bath. Their intimacy was their great charm, and possibly also the secret of their success.

All this has changed. Theatres passed next through the music-hall period. In principle the music hall was not very different from the old theatre; it still had three or four galleries, the pit became the stalls and the upper galleries were enlarged for the mob, and that threw all the balance of the tiers of boxes out of gear, and left the ceiling suspended, as it were, on nothing. It also complicated the entrances.

A more recent development is of American origin, and is a direct result of the influence of the cinema. It takes the form of two huge floors, as big as possible and as wide as possible, and no boxes. No vestige of the old theatre is left except the main ceiling, in the shape of a dome, which was once supported by the ring of boxes, but which is now hung by hooks from the roof. The character of the presentation of the screen or stage changed too. Now everything is a "blackout." The audience and the stage are poles apart, and the stage is as small as possible to spare room for seating. Perhaps we could recapture something of the best that was in the old theatre. Perhaps it has gone for good. Will a return to a modified old theatre, as Charles Urban has done in New York, with several small balconies or tiers and a smaller auditorium, help? Recently built theatres, like the Cambridge in London and the Shakespeare Memorial Theatre at Stratford-on-Avon, approach this development.

How will this affect the cinema?—for the day has come when cinemas and theatres must be interchangeable as buildings. It has been proved that the enormous cinema pays less in proportion as it is enlarged. The cost is prohibitive. The site alone swallows up a vast amount; and as it can only be filled on Saturday night,

DESIGN IN PUBLIC BUILDINGS

during the rest of the week it is often a problem to fill it and make it look busy. We may come, shortly, back to buildings on a small scale, to comparatively small theatres of 1,000 to 1,500 capacity for those intimate plays which cannot be played in huge houses. (You cannot see the faces of the actors in a 3,500 house.)

The interior decoration will also be less overpowering when the scale of the theatre is reduced. Decorative treatment is returning to sanity after the domed and coloured and gilded Italian-cum-New York Renaissance, and the Spanish Patio blue-sky trellised and vined interiors of Los Angeles. All that stuff will go—some of it is going: and we shall have plain colours and simple walls, leaving the audience to give the animation and the stage to be the centre of interest, as it should be. In so many of these over-elaborate interiors the stage was a woeful anti-climax to all but the best plays.

At one time the elaborate gilded carving which used to be found on the front of balconies and galleries and on the proscenium was allied with function. It has been suggested that Mozart's operas were really better heard, acoustically, by being played in buildings that were smothered inside with such ornate decorative work. That was chiefly because all the floriated carving on the front of boxes and galleries was of wood and the material was resonant. A great deal of research and progress still has to be made in acoustical study, in "Talkies" particularly, where the interior design needs to be on very different lines from the earlier cinemas.

Many of the worst faults of the earlier theatres and cinemas, the bad seating and the bad sighting, have been overcome. The logical end of the cinema is, of course, three-dimensional projection, and that is being developed from the back of the screen.

This will mean another alteration in the form of the building when it is perfected. Public buildings can never be static if their function is influenced by scientific inventions. The laboratory must inevitably dictate to the drawing office.

Take hospitals—they are in a state of constant improvement. Even so, they are seldom efficient or economic. As buildings, hospitals get out of date quicker than any other. They must keep abreast of changes. They are always the last word, and yet they are always out of date. To-day they are due for a complete re-shuffle, to align them with new methods of construction and sanitation, new forms of glass, like the type that admits natural ultra-violet radiation, new ideas in medical technique—all these things have focussed into a demand for a new type of building. Any hospital built to-day should have to face internal reconstruction every twenty years. What is needed is permanent floors and structural bones, and movable partitions which can keep pace with changes in medical technique. That suggests a vertical concentration of accommodation. At the moment there is a tremendously wasteful horizontal expansion about most hospitals. They sprawl over unnecessary acres and their design breeds endless miles of corridors. An American doctor once asked whether the staff was provided with roller skates in English hospitals. But the system of vertical concentration is coming to England. It has already started. In the Birmingham Hospital Centre, designed by H. V. Lanchester and T. A. Lodge, which has 800 beds, there is a six- or seven-story centre block, comprising wards and all patients' rooms. There is a large one-floor out-patients' department, a separate block for a medical school and a separate block for administration.

Between £800 and £1,200 per bed is what a new general hospital should cost if properly planned, the difference being the cost of a greater or less degree of equipment.

A new maternity building, recently built at Oxford by Mr. Stanley Hamp, is complete with all services, etc., including kitchen, staff rooms, nurses' quarters; cost, approximately £800 per bed.

The advance in medical science and the special departments necessary for a modern hospital necessitates many of the older hospitals

DESIGN IN PUBLIC BUILDINGS

in the country being remodelled and brought up to date. This in some cases amounts almost to a surgical operation and is often a more difficult and more costly matter than the planning of an entirely new hospital.

The chief task of the hospital designer is to bring about a reduction in the cost of construction. This can be done by vertical concentration. In the number of floors, six or seven stories is perhaps the limit in England. In the U.S.A. everything is contained in a twenty-story block: reception, administration, operating theatres and wards. But then in America the system of heating and ventilation is different. There is not the insistence on cross-ventilation in America: the climate forbids it. Here the medical profession clings to natural ventilation, although the general tendency to change over to single- or double-bed wards would automatically kill cross-ventilation. It might well be asked why cross-ventilation should be considered good for the non-paying patients of hospitals and bad for those who pay? Perhaps cross-ventilation has become an article of faith rather than a reasonable need.

If hospitals begin to take a vertical form we may expect more compact planning, less wasted space. Wasted space means hours of wasted work for the staff, floor space and wall space that should not be there at all.

Most hospitals are crammed with costly gadgets, so-called hygienic rounded corners, tiled walls, windows of eccentric construction, stoves with underground flues, and isolated lavatory towers, which mean long walks through wards for the patients—all these dodges and limitations are the accumulations of architectural tradition.

There are so many old hospitals in which medical students are brought up that professional thinking is hampered by the impression made by the surroundings of the years of training. Some new blood is now being introduced, and maybe the cobwebs will be brushed away. Medical men can encourage cheaper hospitals when they are more enterprising and less traditional. There will be a tendency to encourage small general wards, twelve beds or less instead of twenty, and central sanitary blocks instead of detached towers at the ends of wards. There will be private wards too, with sanitary lobbies next to the corridor.

A really fine modern treatment of hospital design is in the Freemasons' Hospital at Ravenscourt Park, designed by Sir John Burnet, Tait and Lorne. It has no large wards, only a series of semi-private wards with four beds. About 50 per cent of the accommodation is single-bed wards. In the Sutton Hospital, designed by W. A. Pite and Fairweather, the accommodation is about 50 per cent single-bed wards and 50 per cent large wards. In some new hospitals, the big wards are now subdivided with glazed screens.

The most economical size for a general hospital is not less than 600 to 800 beds. As mentioned previously, hospitals cost from £800 to £1,200 per bed. Incidentally a first-class hotel with a bath to every bedroom costs £1,000 per room. The cost in hospitals is swallowed up by special equipment, operating theatres, X-ray apparatus, administrative quarters. A 400- to 600-bed hospital costs about £500,000 to erect, apart from the site. But the hospital equipment for X-ray and other special departments is more than sufficient to meet the needs of one hospital, and could accommodate four or five, with the same expenditure of capital outlay. Actually, hospitals in London and other big cities would benefit from amalgamation. The sites of two or three of the West End London hospitals are worth together at least £1,000,000, sufficient to build a combined 600-bed hospital and leave something over for endowment.

That would at least absolve them from the need of covering the outside of their buildings with disfiguring posters, which, although one sympathizes with the object, are an outrage upon the amenities of any self-respecting city, and which, if displayed in a field near, shall we say Oxford, would raise a storm of protest.

There is no need to have hospitals right in the sunless heart of London. There should be casualty clearing stations at different points in the city, and perhaps three main hospital areas to serve the whole of Central London. These hospitals could be built in not too solid a method of construction, and rebuilt, or rather rearranged, every twenty years. They should be outside of London in open sites. Vertical towers in quiet surroundings instead of being plunged in a roaring sea of traffic noises.

Outer London suburban hospitals should be amalgamated into three or four large units, and each would be sufficiently large to become medical universities. Out-patients' departments would become Ministry of Health stations and be located in populous districts.

Of course that would mean medical officers and surgeons would have longer distances to travel. Doctors give so much and so generously to hospital work that it might make more demands on their time. But it would give hospitals a chance to be self-supporting. Let them sell the valuable sites in London that hamper them at present, and then remove, amalgamate, and concentrate vertically. Concentrated planning does not make any more tax on the available fresh air, and buildings planned like this are cheaper to build, they save long corridors, cover less ground and allow ground for recreation. Hospitals already are attempting to earn incomes for themselves out of private wards.

The L.C.C. is now the largest hospital authority, and should take the lead in concentration.

There is, perhaps, a human prejudice against the use of temporary structures. But many hospitals during and after the war were built of timber and served their purpose well enough. They could be burned down when out of date, although I would not advocate timber now.

Numbers of buildings are out of date about ten minutes after the foundation stone is laid. That will continue to be so, as long as people are working in laboratories, finding out new things, altering scholastic and medical technique. It is certain that the architect of the future who aspires to the design of public buildings will have to have a highly specialized education.

CHAPTER VIII

THE DESIGN OF THE STREET
by
FRANK PICK
President, Design and Industries Association

CHAPTER VIII

THE DESIGN OF THE STREET

In considering the design of the street its functions should be examined, and the first function of the town or city street of to-day is its use for traffic of every kind.

It is often hard to foresee what traffic will use a street, but a clear distinction can be drawn between local and through traffic. Local traffic, or traffic which concerns the street itself, and maybe adjacent streets, cannot be heavy. It may, of course, wait about and be obstructive. It may, indeed, consist almost wholly of vehicles whose owners think of streets as potential parking places. But through traffic is quite different in character. It uses the street to get from place to place. It may grow to great volume; and its volume has no relation to the requirements of the street itself. For a local street there need be two lines of traffic only, one up and one down line. Under present-day town-planning schemes there can be a close, or *cul-de-sac*, making provision for one line of traffic only, provided it is not more than about 400 feet long. It needs a space wherein to turn round at the end, otherwise what would be wanted for roadway can be added to the gardens. It is especially suitable for quiet residential purposes. A length of 400 feet limits the volume of traffic, ensures complete visibility from each end, which prevents any great obstruction, and secures easy access for fire engines, to name the principal reasons.

The character and quantity of traffic as we know it must affect the character and the proportions of the street. A line of moving traffic wants 10 feet in width, so that a two-line street or a minimum street for movement must be 20 feet wide in the carriage-way. This does for all by-streets or side streets. Even 16 feet is accepted. If, then, there must be provision for standing vehicles at the sides of the street, two more lines are wanted, but these need only be 8 feet wide—8 feet for standing, 10 feet for safe movement, to allow for the fact that even the most sober of us do not always move in straight lines. So a usual street width is 36 feet, allowing for four lines of traffic, two normally for waiting or for slow movement and two for free movement. If islands are to be added in the middle of the street, 4 feet more is wanted; then the street must be accommodated by pavements. These are additional, and 10 to 12 feet is desirable on each side. (Unfortunately footpaths are often narrower than that.) So we arrive, if we adequately meet these requirements, at a 60-foot street. Every good street should be not less than this width between buildings, even if the whole of the space is not used for traffic purposes.

This allows a reasonable margin for growth in the volume of traffic. It represents the ideal width for a local street to accommodate eventually four lines of traffic. Where there is through traffic, more lines must be added. It may be necessary to have three or even four lines moving in each direction. But as soon as a street gets wider than four lines of traffic altogether, then the proper course is to begin to divide the street up into sections and to have the lines of traffic in one direction separated by a continuous central island, except for crossing places, from the lines of traffic moving in the opposite direction. If it should be a wide road with heavy through traffic, like the new arterial roads, it is convenient to have laybys for local traffic separated from the lanes for through traffic, and so width is added to width until the 120-foot wide street is reached, which appears to be the limit. This street can be refreshed and made enduringly pleasant by trees upon its islands, so that it never looks the bare and naked place which it promises to be when it is first laid out. Actually, when streets get as wide as this they cease almost to be streets in any real sense of the word, unless they are broken up into strips.

THE DESIGN OF THE STREET

It is when street meets street that trouble begins. But a street and the junctions and intersections of streets must be designed.

A line of traffic moving at fifteen miles per hour continuously can deal with about 1,800 vehicles in the hour. At a faster speed fewer vehicles will pass because more space is needed between them for safety. At a lower speed fewer vehicles will pass. Fifteen miles per hour seems to be what is called the "optimum" speed. When one street intersects another street of equal importance, the capacity of both streets is reduced by rather more than half. It cuts to waste. With all the hazards of intersecting traffic a line of traffic will not usually accommodate more than 600 to 700 vehicles per hour. The defect of London is that there are too many streets. The building sites are too small, and the streets, being many, are poor and unsatisfactory. They really represent a great deal of useless space. A mediaeval lay-out is perpetuated, when whole areas of streets should be replanned to give good building sites. Present street accommodation should be redistributed to afford good wide streets throughout, though fewer of them. On Manhattan Island, in New York, with a gridiron plan, there are eight cross-streets to the mile, and this is now thought to be too many. If New York were to be replanned it is likely that the number would be reduced to six.

Of course streets are widened out in places with the idea of speeding up traffic, and giving fast vehicles a chance to overtake slow ones. This gives us squares and circuses. By setting aside more space and keeping all the traffic moving slowly at the intersection, the capacity of the converging streets is fairly well maintained, but it is maintained at a cost in time and speed to the traffic, and at considerable cost for the room needed to manœuvre, and to weave the vehicles in and out. Because we keep moving we forget that there may be as much waste in slow movement as in first a stop and then quick movement. Maybe it is fortunate that it is an expensive remedy in the centre of cities, for upon the outskirts of cities, where new roads are being built, there are too many roundabouts, some of them quite small in size and therefore merely irritating to traffic. They introduce complications and temptations into driving, slow up traffic, often for no needful purpose, and certainly do not help the pedestrian. For instance, there are arterial roads with two, three and four roundabouts within a mile, which simply make the remedy a worse nuisance than the disease, for applied over and over again it is really no remedy. It seems almost farcical to build these straight clear roads and then deliberately to block them. What is our aim? Free movement or controlled movement? Aim must govern plan.

The problem is really one for signal lights, especially now that signal lights can be obtained which are automatically operated by approaching traffic. Signal lights are precise and definite and the pedestrian gets his turn with the vehicles, which he never does in a circus or roundabout.

Signals are part of the equipment of the modern street. As they do not suffer from being tied to traditional prototypes, they are perhaps the most efficient part of a street equipment. Far in advance of lighting, for example. There is hardly a street in the country which is really well lit by street lamps alone. All streets rely upon lighting from shops and from those additional lamps provided by traders for the sake of advertisement. Shops and stores take on the job of illuminating the street in front of them.

Well-planned lighting should be fairly uniform for safety, not a succession of bright and gloomy spaces, as so often happens. There must be no place for lovers under the lamp itself, for example. No dark islands in the middle with the lamp high above them, which are traps for vehicles. It is necessary to think where the shadows will fall, for it is the shadows which are dangerous.

Low lamps at the sides of the street cast moving shadows from the vehicles on the centre. This is bad. Low lamps in the centre of

THE DESIGN OF THE STREET

the street cast shadows of moving vehicles upon the pavements. That is worse.

Lamps should be high up above the traffic, 20 feet or so, and over the roadway. They should illuminate downwards to give as uniform a distribution of light over the whole surface of the road as possible. The Embankment has just been relit to a fair standard. The Croydon by-pass near the Aerodrome is another instance.

The idea of illuminated kerb stones, which is sometimes advanced as a solution to street-lighting problems, is not practical in its results. The kerb stones would be illuminated, not the street.

The well-designed street must accommodate for lighting purposes posts and suspension wires. There is no need for such scaffolding to be untidy. Really well-designed suspension is beautiful as a spider's web is beautiful, because it is fit for its purpose. It shows the strain and how it is met. The overhead equipment of an electric railway or tramway deserves admiration, where it is well done. Untidiness mainly comes from adding and patching.

But think what gets into a street.

The Post Office comes along and drops in pillar boxes and telephone call boxes.

The Fire Brigade comes along and puts down fire alarms.

The Local Surveyor discovers that a sandbin or two are desirable, or that there must be a dustbin to take street refuse.

The Local Council is stimulated to think of litter baskets and hangs them on the lamp posts.

The local transport undertakings want direction signs to their stations, or stopping posts for their omnibuses or tramways. These are all added.

Then there is a cab rank or a parking place to be set out and marked, or there is to be a signed and marked crossing for the assistance of pedestrians. For all these notices are required. (How averse we seem to using bare symbols.)

And so the notices take the form of regulations printed in full on a board and hung upon a lamp post so high that they are hard to read even with good eyes.

Then the electricity supply undertaking may put down its switchboxes.

There may be signal lights and posts required for the control of traffic.

All sorts of things are dumped into the street without order and without planning by a whole lot of public utility undertakings or departments of local authorities, until the street can become unbearably untidy. All this casual introduction of street equipment is just absence of design. Every street should be deliberately planned from the start, with all its equipment properly located and co-ordinated, for all these things may be wanted.

One of the things that complicate the surface of the street is the frequency of refuge islands. They take up so much of the space of the street, destroying its capacity. They are not placed upon any system. There is no plan for pedestrians crossing a street. The pedestrian wants as much care and direction as the vehicle. More, in fact. Designing streets to suit pedestrians has not yet been started. In our present topsy-turvydom pedestrians spend their time trying to hamper vehicles rather than to assist themselves. In open spaces islands often flourish as a complete archipelago, as, for instance, in Trafalgar Square, where they are a snare to pedestrians. In order to be safe they must be at least 4 feet wide, for they have to carry lamp standards and bollards (the protection posts placed at each end of an island) and give a clearance of 1 foot 6 inches on each side. At first bollards were not illuminated. This was a cause of accident with our poor street lighting. Now the bollards are being illuminated. Every local authority has its own design, its own pattern, generally bad. London illustrates perhaps more profusely than most cities the

PLATE 32
THE DESIGN OF THE STREET

A new shopping centre. The Ilford By-pass. A main artery of traffic.

An old shopping centre. The Pantiles, Tunbridge Wells. A promenade for foot passengers only.

PLATE 33
THE DESIGN OF THE STREET

A street in Pompeii. The stones served as stepping-stones in wet and muddy weather. They also served to check the speed of chariots passing down the street.

A tramway station in a street at Manor House, Finsbury Park, London. There are stairways at the farther end leading down to the Underground Railway station, so that the whole affords a convenient interchange for passengers.

PLATE 34
THE DESIGN OF THE STREET

A CONTRAST

Between a signpost designed to carry its many messages, and a lamp-post that has had message after message hung on it without design at all. The equipment of the street is always overlooked, and then after the street is finished it is added bit by bit without co-ordination—sandbins, fire-alarms, telephone booths, islands, lamps, notices, etc.—with the result that the street becomes untidy.

PLATE 35

THE DESIGN OF THE STREET

A concrete lamp-post in the centre of a traffic circus at the Helder (Holland).

A standard street bollard in Copenhagen, painted grey with bright yellow volutes. Contrast our miscellaneous patterns.

THE DESIGN OF THE STREET

shocking contraptions which can be put into a roadway. Yet the design of a bollard is a simple matter. It needs a single lamp in the cover; no light should be visible from the street, but the light from the lamp in the cover should be reflected from properly designed surfaces, constituting the bollard itself, always of a uniform height. It is a simple scientific problem for the lighting experts, and when solved, as it has been in Copenhagen, or in Berlin, provides a neat and pleasing object which adds beauty to the road at night. It is not exaggerating to say that almost all the designs now in use in London are unsightly and their deplorable variety may be attributed to the fact that our local authorities, even when socialistic in politics, are confirmed individualists in design—and London has many local authorities.

A street implies that it is framed by buildings. The buildings may rise up like walls so that the angle of light becomes important. Forty-five degrees is the usually accepted angle. It is an unhappy convention because it means that the street is in section a square, the buildings being as tall, roughly, as the street is wide. A street would look much better if the buildings were higher than the width of the street, or if the street were wider than the height of the buildings.

The sides of the street were made horizontally articulate first by the architects of the Italian Renaissance.

Nash's Regent Street was an example of such horizontal articulation, with the height of the buildings limited, so that the street was well-lit and sunny. To turn what used to be sunny shopping streets into dark lanes is a modern tendency that makes many cities gloomy. Amsterdam looks like a prison because many of its new streets are lined with blocks of flats four or five stories high, rising sheer from the pavement, so that you have 60 feet of building with 60 feet of roadway.

Naturally the buildings that line any street are controlled by the peculiarities of the building by-laws that restrict their dimensions.

By-laws are useful, but close observance of them tends to create monotony of effect. Nothing picturesque in street architecture ever grew out of by-laws. By-law streets always look unnatural. Berlin maintains a uniform cornice line for its streets. It is an advantage, but it destroys character. Rigid horizontal unity derived from unbroken lines is as different from the fluid horizontal relationships of a great block of buildings as a five-finger exercise is different from a Beethoven symphony.

Some provision should be made for the special treatment of certain buildings as points of emphasis in the street side. In London, subject to rights of light, a building may be 80 feet from the street to the cornice, and then have two more stories of indeterminate height built upon this, so long as they are set back in the roof; thus attics are encouraged. If the rules were rigidly observed, the streets would lack beauty. It is not at all clear that there should even be a uniform frontage line. There might with advantage be some variety in the frontage line for the sake of effect. In this judicious variation of frontage lies the charm of an old-world street. But it is clear that the buildings should be continuous on both sides of the street. Broken lines of buildings can never look really well. But they will always be broken while we design our streets piecemeal upon an individualistic basis.

One of the faults of London is that we have too much freedom. Anyone may buy a plot of ground and build upon it whatever he likes, subject only to the by-laws, to building lines, to rights of light, and so forth. But all the regulations and restrictions cannot produce beauty. They cannot even guarantee immunity from ugliness. As a result, with all the styles of the past at our disposal, our streets are a hotch-potch of buildings. Look at almost any street and see the squalid confusion that is being made.

Often there is no agreement upon such simple things as the level of roofs or floors, or the alignment of windows. There is certainly no agreement as to style or treatment or even material. Everywhere there are ill manners, especially in a business street. The buildings try to shout each other down. Commercialized individualism is a bad patron of architecture. Commercialized individualism must inevitably destroy civic responsibilities in street designs. Only where a large area of ground is in one ownership are streets built to a considered plan. This has happened in Regent Street, and while there may be a debate as to whether the new Regent Street is worse than the old, there can be no debate that Regent Street is a unit, and much better than the streets which are built on a piecemeal, individualistic and competitive basis.

Kingsway is a fairly good street, but only by reason of the restrictions placed upon its buildings by the London County Council, and it is spoiled by its two churches, which are out of scale with the other buildings. It is necessary to go to the eighteenth century to find streets where all the buildings are related one to another and built to scale. Only in the eighteenth and early nineteenth century was there planning and building of streets on a grand scale. Fortunately, in Bloomsbury or in Bath, in Brighton, or Hove, or Cheltenham, or Tenby, and some of our older towns, good examples remain to us.

Streets are not improved by the higgledy-piggledy of a suburb. They must have *shape, definition* and *direction*. Traffic demands that a street should have direction and not meander to please some town-planners. Civilization, which stands for urban amenities, demands that a street should have shape and definition. It is no good building houses where one house looks on to the side or back of another. It is not a satisfactory prospect. The hit and miss of villa and garden can be just as monotonous as a row of houses. After all, while people are all alike as people, yet they differ as persons. So should their houses, in reason, and the differences should be civilized, not comic antiquarian differences. Some buildings behave like boors or clowns.

Others get into fancy dress and pretend that they are subjects of good Queen Bess, or sober Queen Anne.

The character of the street should be reflected in its architecture and in its design generally. There are residential streets which should be quiet and secluded. *Culs-de-sac*, if possible, adapted for local traffic only and paved for quietness. There are streets dedicated to certain professions, like Harley Street for the doctors. These must be wider streets, to allow for callers with waiting cars, but they should still be backwaters from the main flow of traffic.

Then there are industrial or factory streets, which must be wider again to allow for the manœuvring of vehicles for loading and unloading, though this should not necessarily take place upon the street, but in areas provided off the street. Here the road should be built up with solid foundations and with a hard crust, to take the wear, as the streets will be used for heavy traffic.

A shopping street should be a street in which it is possible to loiter and gossip and gaze at the shop windows without being jostled. Wide footwalks are wanted. There is no reason why a shopping street should not have a promenade up the middle like the Unter den Linden in Berlin. This divides the street really in two, one up and one down, but it makes passing across the street easy. Where a shopping street is a wide street with through traffic it is almost impossible to pass from side to side to look at the shops in comfort.

Then there are pleasure streets like Shaftesbury Avenue in London or the Deeperstrasse in Hamburg. By day and night they should be bright and gay and with room to accommodate crowds of folks when the theatres and cafés leave. There seems no reason why illuminated signs should be restricted in such streets. All that is necessary is that the buildings should be designed to carry them properly, rather than have illuminated signs hung all over the façades. It is again only a matter of design. Illuminated signs at Piccadilly Circus would be all right, as Piccadilly Circus is a pleasure centre, but illuminated signs at Trafalgar Square are all wrong because Trafalgar Square is largely surrounded by official buildings. The mere invasion of the square on one side is just unsightly and rude, an example of civic irresponsibility.

Streets which are in the nature of processional ways require laying out, that people may see the show. They might be terraced off to afford a good view to spectators. The Mall, for instance, might have been made a much more useful street for processions if a little more thought had been given to it. Whitehall might have had its pavements raised by two or three broad steps, like old Kensington High Street. Certainly there is more road space than is required for the traffic, through the major portion of its length—which, by the way, is cumbered with indifferent statues—for the volume of traffic which passes through Whitehall is governed by the bottle-neck at Trafalgar Square.

At present, owing largely to lack of design, most streets have too many different jobs to do. But we are already thinking of the street of the future. As the congestion of traffic gets worse, it will be necessary to lift the centre of London up into the air about 15 to 20 feet, and double deck all streets, providing for the goods and heavy service vehicles and so forth to use the lower level and the passenger vehicles to use the upper level. The shop fronts would give on to the first floor. The vehicles would run along a kind of continuous promenade across the front of the buildings. It would be possible to look over and see in the middle the heavy commercial traffic moving laboriously and continuously. Even before 1914 the Grand Central Station, New York, was built like that. There is a drive all round the building above the main street level, connected by over-bridges across the main street, with lateral streets to avoid conflict of traffic.

This is one remedy. Another would be for vast new buildings to enclose a series of shops; streets within streets, so to speak. This would restore intimacy to our shopping. It would give far more space for display under adequate control. A series of small shops, built together in layers, like a honeycomb, would be most attractive, with a proper scheme of lighting and ventilating for the connecting ways. Shops within shops, that is the secret of the really attractive store. No one has yet thought this out in detail.

Fancy soon runs a long way in these speculations, and it is best to stop before it becomes lost in the distance. If we take pains with design, then the street may become a work of art. We never think of a street as a work of art, and the present state of our streets is the result, not of our neglect to think of them as works of art, but of our neglect to think of them at all.

CHAPTER IX

DESIGN IN THE COUNTRYSIDE AND THE TOWN

by

E. MAXWELL FRY, B.ARCH., A.R.I.B.A.

PLATE 36
DESIGN IN THE COUNTRYSIDE AND THE TOWN

PLANNING IN THE GRAND MANNER
Actual view of Bath—grandeur on a small scale: the really civilized town.

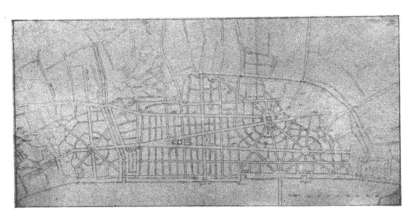

Sir Christopher Wren's plan for London—a great opportunity lost for ever.
(*From the original in the library of All Souls College, Oxford.*)

PLATE 37

DESIGN IN THE COUNTRYSIDE AND THE TOWN

(a) The strangulation of the town — industry and housing in an unholy alliance that is usually shrouded in a smoke pall.

THE AGE OF UNPLANNING—THE NINETEENTH CENTURY AND AFTER

(b) The defence against industrialism — garden city development, or informality run riot, and still no town to live in, and less country to enjoy.

Opposite: An air view of the Hampstead Garden Suburb. The first completely realized garden city project. (*Photo: Aerofilms.*)

PLATE 38
DESIGN IN THE COUNTRYSIDE AND THE TOWN

PLATE 39
DESIGN IN THE COUNTRYSIDE AND THE TOWN

THE COUNTRYSIDE
(*a*) The English countryside as man made it—a pattern of hedges, fields, trees, and separate, closely-built villages

(*b*) The English transformation-scene—the country being swallowed up entire by unplanned housing and cast-iron road-making; where will the town end?

CHAPTER IX

DESIGN IN THE COUNTRYSIDE AND THE TOWN

SINCE very few of us live in towns consciously planned as such the idea of "designing" on a large scale is not one that we can grapple with at once. Used as we are to being provided with every form of public service on tap, we still boggle at the logical consequence that leads us onward toward the consolidation of all services, into a completely planned community structure, a town "designed for living."

It is because we fail to notice the confusion we live in that we set a low estimate upon the nature of design. Fly over any large town in an aeroplane and you may see the awful waste of building that has been scattered over the face of the land, waste that sprawls from the congested centres, rushes along lines of least resistance, halts before obstacles, is deflected, baffled at every point, hardens here into points and knots of congestion, thins there into wasted barren patches. It is like a spreading plague; a blindly directed blight upon the face of the land.

But you cannot, you may say, plan a town, that is, a living organism, as an architect plans a house, which is a thing fixed in time. A town grows like a forest. Yes, it does; like a forest, an ordered and well-controlled forest, but not necessarily like a jungle which in its natural state is a wild conflict of growth.

A town is much more complicated in any case. Its origin may have been simple; but its development, as it acquired fresh functions, became complicated. Towns began as markets. When men built a market centre they created the beginnings of town life, which is artificial. A town is an artificial creation, based primarily upon the need for a centre for the exchange of goods, which reaching a stage of permanence must provide adequate shelter for those who live and trade there.

Shelter is a good word. It covers every form of building: large houses and small; for even in the simplest type of town there is a classification of dwellings. In our scrutiny of a town's functions we have therefore got *Trade*, or let us say *Work*, and *Shelter*. There must be a thread on which these functions are strung, and that connecting thread is the street. There are the three chief elements of a town; *work*, *shelter* and *communications*. To these must be added *Relaxation*. Work, shelter and relaxation, joined together by communications. Those are the vital elements of town planning. Those elements can only produce real civilization if they are assembled by design. They have been in the past; but design does not seem to be understood in England to-day, even allowing that the problem of design in towns is not as simple as it was a century ago.

Let us bear in mind the four fundamentals, for they still operate most strongly, and let us, before applying ourselves to the problem of large-scale design, see what steps are taken in the smaller and more compact problem of designing a house.

First, being a rich man, let us say, you find a site that suits you, being healthy, having communications with a market or shops, and water, gas, electric light and services laid on or made on the premises. Then you count your family, estimating future additions; and you fix on your rooms and plan roughly how your house will work, uniting separate rooms by corridors, placing the kitchen quarters near the dining-room but away from the garden and so on. *A town is planned in the same way.*

Take Welwyn for example. That was planned after the war, not very long ago. Its site was chosen as being on the main line from London, between the railway and the Great North Road. It is a healthy site and in good country. The plan of the future town was laid out on paper, very much as a house plan is. Acting upon our knowledge of the evils of unlimited expansion, its planners limited the future population to 30,000 people. By doing so they were able to say how big the town should be, and they drew a boundary line and said this is where town ends and country begins. When 30,000 people occupy Welwyn it is finished entirely, though before that figure is reached the town would found another separate centre. Having drawn their boundary, they reserved an area to the east of the railway line for the reception of industry, upon which the town was to depend for its existence. The remaining land they divided into districts for living, for shopping, amusement and government, and these they connected with main roads, secondary and service roads. Nor did they forget parks, which are scattered about in small open spaces, in the fine lawns which dignify the centre and in smaller playgrounds. Further, of course, no part of a small town is very far from the country, and Welwyn has its cinemas and theatres and all the intelligent and exciting contacts that a town alone can set down in a rural setting.

Welwyn shows what can be done, but Welwyn was started with a clean slate. It has grown and will continue to grow graciously because it is controlled centrally. Unfortunately, unplanned growth is the common lot of English towns and cities. A hundred years ago England was not an industrial country. Its population was eight million. London had about a million inhabitants. Society was rigidly stratified, rich and landed aristocracy taking the social and cultural lead as by natural right. This society, when it built or designed nearly any mortal thing, had a fine sense of order, which owed its direction and its force to classic or Roman culture. The distinguishing trait of classic culture is breadth, repose, symmetry, a sense of scale—that is to say, the fine adjustment of one part to another—and a fine reticence of expression. What has been justly called the Grand Manner. And the Grand Manner distinguished the eighteenth century in England. What was true of the design of the eighteenth-century houses and furniture, their noble simplicity and dignity, was true also of the towns. Sir Christopher Wren's plan for London would have made of London an architectural masterpiece, and Charles II would himself have had it so. It was the tragic combination of fire and war menace that defeated him—these and the popular clamour for an exact partition of what was left from the burning. Common people had no time, alas, for visions.

It is absolutely safe to say that if Wren's plan for London had been carried out in the seventeenth century we should have no terrible traffic problems in the city to-day; and where men of a like mind laid out the best of Bloomsbury and Mayfair we ask nothing more commodious or noble than the streets and squares they have left to us. In spite of growth and change these designs have lasted well, and their character is the essential character of London.

Into the Georgian age, with its bland adjustments, its stability and harmony, into the midst of this well-ordered though top-heavy society a few earnest inventors threw a bag of the most powerful bombs that ever disturbed the world. The steam engine, the spinning jenny, the factory system, railways and steamships. There was an explosion the like of which the world had not before experienced. It wrecked the entire structure of society and in fifty years exterminated its culture. There grew up a teeming population of factory and mine workers, an eruption of industrial towns. Lancashire, the Black Country, the North-East Coast areas, South Wales, all these dirty, crowded places, which are to-day the seats of unemployment and the subject-matter of slum clearance, became then the happy hunting grounds of a new type of individualism that knew no laws and had no taste. There was planning, but it was a case of how to plan a factory and as many houses as possible into a narrow space. It was "unplanning," condoned by an economic creed that was humorously known as "enlightened self-interest."

DESIGN IN THE COUNTRYSIDE AND THE TOWN

The congestion and the overcrowding which surround us to-day are the result of this accident, which came too quickly—as accidents *do* come—to be digested and organized into useful channels.

But we are not fated to accept the mess we live in as inevitable.

Admittedly it is discouraging to contemplate mile after square mile of densely built-on property of London and Manchester elbowing its smutty sleeves into the surrounding country, and all the horrible sore places in the Midlands where residential and industrial districts intermingle and give a general effect of a Satanic slum. But in time all this can be cleared up.

For a town is an organism. It is always changing and decaying. So long as you allow yourself to believe that you cannot afford to plan, all change adds to stagnation because congestion makes you replace each rotten building with a bigger one, and the next stage is worse than the last. Take slum clearance, for instance. Suppose that in a certain old quarter of a town living conditions have become intolerable. The common course is for the town council to buy a section of that property, to pull it down and to re-erect exactly upon the site block dwellings of four or five stories, throwing the old backyard spaces into a common court. Well, perhaps it is better to do this than nothing at all, and some such dwellings are very pleasant affairs, though usually lacking in imagination. But is the town much better off for this piece of social trepanning?

Only a little better off, for no attempt has been made to take the opportunity slum clearance offers to straighten up the street plan, or to right the lack of balance between built-up area and open space. The thing is piecemeal, and no more planning than face-lifting is surgery.

But a town is not all houses: there are factories and railways. These things should be planned. So far, they have just occurred. Just how casually such things occur is illustrated by a map of Edinburgh before and after the coming of the railway.

Edinburgh was beautifully planned for living. It is still one of the finest cities in Europe. No doubt the railway, when it came, was beautifully planned for transport; but nevertheless that railway made an awful hash of the road plan, despite the fact that both types of planning were made to serve the same society.

Nor are factories an end in themselves so important that no one must question their location. Houses, roads, railways and factories—all serve society. They are all susceptible to planning with the object of making a beautiful town fit for everyone to live in.

Town planning is still developing. It arose out of the terrible conditions of towns in the last century, conditions that were bad enough to stifle the early reformer's hope of doing anything effective at the centre; therefore they turned their eyes away towards the outskirts, and using a romantic notion of architecture then prevalent, worked out the system of garden city or suburb development upon which the movement has up to now been based. The industrial town was crowded to a degree, forty houses to the acre being common. The new system went to the other extreme and gave every house a garden, front, side and back, and spread them at twelve, eight and less to the single acre. The desire for a garden is natural, and a house surrounded by garden is more human than the crowded industrial town houses with their grimy backyards. It does at least provide light, air, trees, flowers and some connection with the soil. But the system developed on the fringes of our artificially created towns, already too widely spread, and it was not town building at all. It made the mistake of pretending that sedentary workers and artisans can live like rural workers. In insensitive hands it produces little houses littered endlessly along winding concrete roads, interminable lines of machine-made Tudor or bijou baronial halls, extending north, south, east and west out of every sizable city in the country. For this the speculative builder is more than half to blame. But even in the best of housing schemes the thing does not work, because the joys and the beauties of the town are those of close congregation, and no architect can plan these little cottage units to produce the grand effects of the town—they remain cottages or villas—where indeed they do not degenerate into bungalows.

The term "garden city" is often misused to-day. In Hampstead Garden Suburb you will find the garden city idea at its best, for there the romantic material of cottage type building is planned into one coherent whole centred upon the "square of the two churches" which dominates the suburb.

Letchworth was the first great experiment of town building with garden city ideals, and as an experiment it must be regarded as highly successful. It worked out the principles of what is called "zoning"; that is, the division of the area of a town into districts serving separate purposes, such as industry, business, local government, residence, recreation. It confirmed the existence of those vital elements: WORK, SHELTER, RELAXATION and COMMUNICATION. In the next experiment that followed at Welwyn some attempt was made at the centre to revive purely urban architecture by the arrangement of continuous street façades, with the garden element subordinated to the purpose of the general design. In years to come, when the central portions of Welwyn are completed, its present state of unbalance may be rectified, and we shall be able to see how far the garden city reformers carried us towards the long-forgotten art of civic design.

Meanwhile, we go on adding thousands of semi-detached villas until it seems the town, or this headless monster which is neither town nor country, will never end. We adorn our fine, tree-planted arterial roads with small but pretentious, ignorant and vacant villas, copies at fifth hand of houses designed by architects twenty-five years ago. This folly we commit partly because the doctrine of "enlightened self-interest" still upholds the divine right of muddle making, and partly because town planning has become a sort of science divided into watertight compartments. The people who lay out streets to-day are engineers and surveyors, who do not realize that a street is nothing until you erect buildings along its borders. The relation of these buildings to the street gives it nearly the only form and character it possesses, and if the houses are as silly as they are on the Great West Road, and as vulgar as some of the childish factories that form this ignoble gateway to Western London, then that street, no matter how wide or how costly it may be, will be as silly and as vulgar. What a tragedy it is, to plan with no end in view and with no architecture! That infamous thoroughfare, the Great West Road, was laid out on the best models, with grass verges, trees and fixed building lines. It might have become a noble entry into London. As it was, control ceased with the road itself and its borders were allowed to receive the backwash of the garden city movement, and all the worst sentimentalities of uncultured commercialdom.

Meanwhile we have these great industrial towns and sprawling London to get into some form of order, and time presses, while the drain of sustaining the continuous waste of life and money continues. Every part of London shows where the machine is broken down, clogged, rusty, decayed. There are places where town planning could make possible slum clearance on a scale that would free acres of open space for the enjoyment of people now crawling along foul alleys and rotting in underground rooms. There are places where roads that end in despair could be carried triumphantly onwards, bearing rich loads of traffic to the ease of other parts. There are miles of two-storied cottages, interminably spreading without relief of pleasant green, rotten inside and out; ready for the rebuilding of a more humane London. We have got to get into our heads the idea that by planning we shall be saved from the waste of life and money that flows from us while we procrastinate. We failed to solve the problem of Charing Cross Bridge because we had no plan and did not know what to do. We are failing to clear the slums to any good and lasting purpose.

DESIGN IN THE COUNTRYSIDE AND THE TOWN

In London the L.C.C., a single body charged with the care of London, is unfortunately a parent body with a family of local boroughs; and its town plan is no more than the scattered pieces of a jigsaw puzzle strewn about the nursery floor of London. *There is no plan of London. No plan and no survey*; only a mass of information which no one knows how to use.

There must be an enormous amount of information docketed away in Government Departments and such places, but because there is no survey and no official call for these valuable facts they are largely unobtainable. In sober truth, surveys of London and the other great provincial cities, and a national survey to follow, are as much needed as a reduction in rates, for without them we cannot plan—unless, of course, we are content to be *amateur* planners to the end.

People get very confused about town planning and imagine that it must necessarily mean building and spending, whereas it nearly always stands for saving, and very often prohibits building. A wise landowner thinks years ahead, deciding how best he can conserve his resources; where he will build in the future; what it is wise to spend and what to save. And this he does, weighing the possible benefits of each action with the welfare of his whole estate. Town planning is the wisdom of the wise landowner applied to towns and country, and eventually to the whole country. The conduct of town planning lies with those elected to govern nationally or locally. But the results of wise planning are to everybody's benefit.

The more everyone knows about planning, the better the planning. The popular demand for the preservation of the countryside has stirred the authorities to some action, so that the nation has been presented with parks and reservations all over the country.

The countryside itself represents two separate problems. First of all, agriculture is an industry, and an industry which is not confined within four walls, but spreads over the whole face of the land. The primary use of the countryside is for the industry of agriculture, which through the splitting up of the large farming estates and the slump in prices is now passing through a very difficult period. The spread of motor transport and all its by-products in petrol pumps, and shacks and other roadside equipment—variously disgusting. That is the second problem; but had the great estates remained intact it might have remained a minor problem. Once land is split up and sold in small pieces all the old control vanishes and muddle takes its place. Planning in the country, if it is to mean anything at all, must first of all keep the land free for agriculture, and guide building into groups and towards the towns. The countryside stands first of all for agriculture, for work.

But the townsman must have his relaxation and transport must have its roads, and this is the second problem: reservation and preservation. The artificiality of town life and its stress and excitement require antidotes. When townsmen go into the country they seek the "wilds" or look for beauty. Places like the Lake District or North Wales are the wastelands of agriculture, but the holiday grounds of the people. They should be national parks. They are only lightly used by agriculture, and there is no reason at all why we should not reserve them for their present purposes, such as sheep rearing, occasional mining and so on, with reasonable right of access for the public. But primarily these national parks should be kept in their semi-wild state. Large areas must be preserved from haphazard building and every form of defacement. Then, besides the Lakes, there are the grass Downs of the South, Dartmoor, the Broads and other places. There should be natural reservations within reach of every large centre of population. They should, in fact, be planned for the greatest usefulness. There are the commons and footpaths wandering through the fields away from motor traffic and crowds. None of these good things are so secure that we need not include them in schemes of country planning. As for the petrol stations and roadside tea-shop abominations—they are the scaffolding, as it were, put up in a transitional period on which in time something more orderly and civilized and good-looking will be built.

Even now the idea of planning is definitely gaining ground. In the Middle Ages people died like flies from diseases and plague, until at last they came to see that a few people were talking sense, whereupon they ceased to pollute their drinking water. In the same way people may come to see that Town Planning is sense, and then we shall set about the reordering of England.

CHAPTER X

THE MEANING AND PURPOSE OF DESIGN

by

FRANK PICK

President of the Design and Industries Association

CHAPTER X

THE MEANING AND PURPOSE OF DESIGN

In the foregoing essays everyday surroundings and the things of everyday use have been criticized, perhaps with a severity that some readers may consider mordant. It may be felt that critics must be unpleasant folk, seeing the worst rather than the best of things. But this is not quite fair, for a critic is one who first understands and then judges, and he should know as well how to appreciate as how to condemn.

Although I am President of the Design and Industries Association, I can make no claim to be a qualified critic. I am still trying to understand things myself, but I have already ceased to be afraid to say what I think about things, even at the risk of sometimes saying something wrong, which is better than saying nothing at all for fear of making a mistake. If the things which we use are to be improved and bettered it will be only by continued and, as far as possible, by well-informed criticism.

It is something of an effort to look at the things that surround us with our minds as well as with our eyes. We so soon accept them unthinkingly. That is one excuse for a good holiday every year. It uproots us, and that opens our eyes to see things afresh. On our return home, for a day or two at least, we are inclined to question the features of our environment, but habit soon returns us to our comfortable grooves of thought. If we cannot keep a healthy, questioning spirit in any other way, then we must make a business of it. We shall find it an amusing, interesting and instructive business.

If you take this business seriously, then you may become missionaries throughout the length and breadth of the land for the conversion of things from bad to good, or from good to better, or from better to best, if that is possible. You must first startle your shopkeeper by asking questions about fitness; the sort of questions that are asked and answered in this book. The retail shopkeeper may then startle the manufacturer by passing them on, and the manufacturer will be compelled to look about for designers skilled and competent to embody the answers which you want in the things themselves. The movement starts with you, reader, the consumer or user, for you represent that large, powerful and influential force called "the purchasing public."

Now this matter of design is not abstruse at all. By nature we are all designers. It is only a matter of using our brains, of bringing thought to bear upon the making and fashioning of things for everyday use. It is putting as much brains into making a pot or a pan as a wireless set. When men were making the first pots and pans they had to think about them, but now we are so clever we can make pots and pans on a mass production basis with scarcely any thought at all. A wireless set is still an obstacle needing quick minds and nimble fingers, demanding craftsmanship for good results.

But a frying pan, to take a definite problem, is not the simple affair we may imagine. It must be of solid metal which will absorb and hold heat. Thick is better than thin metal. If we ask for a light frying pan we are forgetting its main purpose. The handle must be firmly riveted or welded on, yet the handle must not get hot. It must be of metal shaped to dissipate heat. Design is applied science at this stage. Those of us who use frying pans know how often we scorch our hands, how often the handle wobbles, and we know how much better a pancake is that is cooked quickly and not slowly. Now, is our experience turned to account so that we cannot go into a shop and get an unsatisfactory frying pan? Well, not always, but it ought to be.

Thought in the shaping of things can be analysed very simply.

THE MEANING AND PURPOSE OF DESIGN

First, it means order. We all appreciate symmetry. Cut an ordinary ornament down the middle and the halves are alike. Or have a pair of ornaments, one on each side of the mantelpiece, to match, as we say. This question of balance starts with simple things like handle and spout on either side of a teapot. There should be a proportion or relationship between them. They should let the teapot look as though it were not being weighted up or down one side or the other. It may also be studied in flower vases which have so narrow a base that when filled with tall flowers they are unstable. But perhaps the best illustration is found in chairs. The wooden chair that stood firmly on its four legs has given place to the metal chair which looks as though it had no legs at all and yet is just as stable. In some, the upholstered seat and back look perilously suspended. Such a chair has the added advantage of using its bent metal tubing to give resiliency which the wooden supports were unable to give, but when not in use the seat is not quite flat but slightly tilted upward. This looks odd, but it exists to counteract the weight of the sitter, so that it is level when in use. It may be compared to loading a spring on a motor-car. It shows that the design is adapted to the material and incidentally a true application of fitness in design.

Nowadays, we find symmetry too obvious, and we deliberately set ourselves more difficult problems of balance. We know that masses, which are far from symmetrically disposed, will yet balance round a centre line or point, and so we set ourselves to judge nice and difficult problems of asymmetry, or the opposite to symmetry, and if we are clever enough to judge aright we get more pleasure out of it than out of the obviously symmetrical.

The old balance of symmetry looked at rest. One part definitely offset another part. The new balance looks as though caught at rest just for a moment, as though it would move if the slightest change occurred. It has the attraction of being caught and held just at the right moment. Afternoon tea tables with shelves at odd levels, or stands for cactus pots with little ledges climbing about, as it were, look balanced when they are cleverly arranged. Orderliness and use do not enter into our judgment to any extent. Unluckily, this taste for asymmetry may lead us astray. It may be fitting for an occasional table to slip things into; it is only foolish for a writing table, for example. There you must sit with all things about you in an orderly fashion, which rules out any playful variety of level or arrangement.

Order is a notion which we all try to put into practice. Look in a cupboard and see the tea-set or the dinner-set packed away. The better it is packed away the better it looks packed away. The cocktail cabinet has become a masterpiece of neatness, and so of design in this sense. It holds such a lot with its racked glasses on its doors and its ingenious fitments that economize every bit of its space. It is the same with a fitted wardrobe. If well fitted, it gives a sense of order confirmed by experience, which is an added grace of design, so that we look back with shame upon large, ill-equipped cupboard-like affairs or vast drawers in which things tumble about. This virtue of order in design is the same as the virtue of tidiness in ourselves. We all know the bedroom of the untidy man. Socks on this chair; suit thrown over the back of that; here a shirt and there a collar; the drawers not closed; a cigarette left burning on the mantelpiece with the marks of many scorchings which have gone before. Admittedly all this may make on occasion an attractive human picture, but not a room to live in; and if the outward show is like this, what about the inward, represented by the insides of unseen receptacles? We know this untidiness is not designed. Now look at the furniture that fills your room in the same way. Is it arranged in an orderly manner? Loose furniture is giving place to fitments on a unit pattern. It is a great advance in room design, giving more space in our smaller houses. All we have to ask is whether it is made up of useful parts to hold things properly. Whether any part is idle or ornamental.

If so, what is its value to us? Even the untidy man may blame untidy furniture for his own untidiness.

Thought next means the fulfilment of purpose. We think out how we can do something, and if we think well, we can do it better. Take our knitting, for example. We begin on the plain jobs, "knit one, purl one," and only later take up the "knit one, wool forward, knit two together" patterns. We begin with mufflers and finish with pullovers, or even with our bathing costume; but we buy the ornamental flourish and stitch it on—the fish, or flower, or mermaid that identifies it. One day we shall venture on this decoration. Our knitting, when done, must serve our purpose, or else we give it away to an enemy, aunt, niece or whatnot. We know that stage by stage in our progress we must put thought into it. Habit only comes with practice. So we must see that when we buy things thought is put into them.

This applies to dinner-plates, for example. Dinner-plates should hold a generous helping: the rim rising clear of the sauce or gravy, not too steep to hold the salt and mustard without slipping, a trifle hollow for this; a nice colour not to clash with the food. (A tone of green or blue does not improve the appearance of the meat.) Then we want to know if the plate is clean, which means the pattern or decoration should not sprawl about the centre, hiding any marks which there may be, and even at the edge it should be plain and clear. There is a chance for thought in the choice of a dinner-plate, if it is to fulfil its purpose satisfactorily, which goes beyond the mere pattern on it. Or take an easy chair. Support for the head is often lacking in these modern, low, box-like chairs. There is too much support for the arms in them also, which pin you in and hinder writing or knitting.

But when we come to this test of purpose, all of us are competent forthwith to pronounce judgment. We use these common things, and our purpose is served or it is not served. We should be quite outspoken about it. Fitness for purpose, that is the basis of good design.

We seldom do well anything that is easy and effortless. Presumably we get careless. That is the only reason I can give for the shameful things that can be bought in brass, in white metal, in celluloid and similar materials. Glass is a beautiful material, transparent, capable of taking on beautiful, glowing colours, of reflecting and transmitting light. But it is also a material with which you can do anything. When molten and fluid, it can be blown to a fine bubble and worked into almost any curved shape; or it can be cast into almost any form. When cold and hard it can be cut and polished; it can be etched and ground. It will submit to any indignity, and it has to submit to many. Because it is so accommodating and obliging, it is surely your duty, and privilege even, to protect it from abuse. Examine your glass, and see whether you have done your duty by it. Does your glass recognize its qualities of transparency, of reflecting and transmitting light?

Think of the glass, thick and heavy so that you may deeply score its face all over into diamonds and squares and so forth. Does it really sparkle like a jewel when it stands on the table full of flowers, or when it arrives from the kitchen filled with salad? Should a salad bowl glisten like a jewel? Is it fitting? Is it not presumptuous? Cut glass is rather vulgar. It likes to look richer than it really is.

Think of your mirrors. Glass rolled and smoothed out at great pains to reflect what it sees accurately, and expensively silvered and finished off at the back. One thing they cannot reflect, and that is what is painted on their face, yet paint them you will. It is the contradiction of fitness for purpose. You can score a pattern or figure on a mirror which will pick up the light and add a touch of fantasy to it.

And then your wine goes into glasses. In itself a lovely colour—amber, or gold, or orange; ruby, or crimson, or purple—and you sometimes put it into blueish, greenish or brownish glasses, giving them a fine name like amethyst, sapphire, topaz. It is all too easy, but it really won't do. Can lemon squash even look well in topaz glasses, or orange squash in amethyst?

If a thing is hard to do, we have to think before we do it, and then we are not likely to make a mess of it. Take the wireless set. The chassis is a difficult affair, and we can hardly make a bad job of it. If we do, the wireless set is cast out as useless; it is immediately condemned. Open the case and see the orderly wiring, the grids and screens, the carefully finished valves in their containers; everything in its place, nothing superfluous or merely decorative. Yet it has a pleasantness of design we can all grasp. Close the case again, and what do we see about it? That fret-saw that gave to the mouth of the loudspeaker a rising sun over a billowy sea, or a couple of willows by a stream. It is so easy that thought was wasted on it. Then the handles to lift it by, if it is a portable. Can we get a firm hold of them? Wireless is of the twentieth century if you like, but should it be put in an eighteenth-, seventeenth-, or even sixteenth-century case? The period case! Not all wireless cases are bad, of course, and a word of praise is due to those recent designs which have had regard to modern requirements and modern sensibilities, which are fit—the exterior and the interior being in some sort of harmony.

What is to be said upon this point of ease or difficulty in manufacture falls under two heads: firstly, a respect for material; secondly, a regard for craftsmanship. We are often afraid to recognize cheap material. It is a form of snobbery, and so we attempt to make it look what it is not. We all know the mock marble that comes from paint, or the deal that apes the oak with graining. Plywood has come to our defence, for it may wear a lovely face honestly enough, and modern design must take account of plywood. Even the silver gilt of the rich often seems to me to indicate a lack of real taste. It is only another form of snobbery.

We often fail to recognize the labour of the maker. We have our silver smoothed out and burnished sometimes, even when it is handmade, so that the work of the hand is entirely lost and hidden. And then, "contrariwise" our brass pots, which are spun on a machine, are filled with steel bullets and shaken hard so that they may look as though they had been beaten out with a hammer. Any respect for craftsmanship would stop all this. Try, therefore, to imagine, when you buy, how the thing was made, and look for the natural and healthy signs of labour.

Watch a blacksmith humouring the hot metal, and with heavy strokes now this side, and now that, giving it the shape he wants. Or a carpenter with a gouge or chisel making sure and certain cuts in the wood, so that they have a crispness and sharpness which hold something of beauty always. Then think how different the effect would be if the strokes or cuts were fumbling or tentative. The resultant shape might be the same, but all the surface and texture would have been destroyed and spoilt.

Then by way of contrast watch a machine. Take a rotary planing machine. The rough wood goes in on one side, and on the other side the boards come out of a regular and even thickness with their faces smoothed and true. If there were any play such as the hand might give it would mean the machine was "out of true" and wanted adjustment. The signs of craftsmanship have altered.

When you buy painted pots where the colour has been added by strokes of a brush, seek those signs of directness and sureness which the hand can give. Appreciate the slight variations in the pattern which flow from them. But when you buy pots where the pattern has been printed on or taken from a transfer, seek then the qualities which a machine should give. Be sure the transfer is carefully placed, that its ends join together exactly. You must all have noticed on a plate the break in the pattern where the transfer has not joined exactly. There is a craftsmanship of the hand which is expressed in

THE MEANING AND PURPOSE OF DESIGN

freedom, and there is a craftsmanship of the machine which is expressed in determination, in accuracy.

Design is something more than just using material appropriately, or showing skill in its use to bring out its beauty. It is something more than serving a purpose. It is something more than ensuring tidiness and order. It is also expression. It is the designer saying something for himself. The further we get away from use, the more we must seek expression. Fitness for purpose must transcend the merely practical, and serve a moral and spiritual order as well. There is a moral and spiritual fitness to be satisfied. We know it sure enough when we see it. In our charming country cottages the furniture is mainly for use, but the charm depends upon it being also an expression of the life lived in the cottage. So in our homes we should express our lives. M. Le Corbusier has said "that the house is a machine for living in." That only partly expresses what a house is. A house is also the setting for life. I am not at all sure that we should not help our friends to guess what our employment is, what our interests are, what are our fancies and ideas, when they come into our house. So often we hide the personal note and produce what is a conventional room, which is often without character. There should be no concealment. It is snobbery again in another guise. *Into the void that snobbery makes come the devils of fashion.*

So with the things that go into the house. They should have an expressiveness of their own as far as possible. Wherever they are ornamental or decorated, we must insist on this expressiveness, for it alone can keep the ornament or decoration within reasonable limits. Textiles demand it. Perhaps we reveal ourselves most in our curtains and carpets. Certainly we do in our clothes in spite of convention. The most dangerous things we buy are textiles.

Possibly there is only one right solution of this problem of design. Many of the common things which we use have had their design settled long ago. The face of a clock, for instance. The hands go round, and so the figures should be arranged in a ring. Yet in these days you get clocks which are triangles, squares and oblongs. The hands at one moment are trying to reach the figures on the margin, the next moment they entirely overlap them. There is only one right answer to the design of the face of a clock. Pity the wretched designer forced to improve on the best for the sake of novelty! But it is the consumer's fault. The designer is not naturally so perverse. If the "purchasing public," prompted by the devils of fashion, would not buy them, then they would not be made.

Of course there are things which, luckily, allow of a convenient variety of solution, like the length of a skirt, which may wander from the knee to the ankle, or even down to the floor and trail. But the height of, say, a street bollard is fixed. It is a sign, a warning light, which should always be in the same relative position.

The illustrations I have chosen have been from homely and familiar things that have nothing to do with art or aesthetics, or beauty in itself, and yet they may be artistic or fine or beautiful. When we have to deal with such things, we may feel justified in approaching them along easy, recognizable paths. We can ask whether they are orderly or tidy. We can ask whether they say or disclose anything to us beyond the bare thing itself. We are on fairly safe ground, and if to all these questions we can give anything like a satisfactory answer, then the thing, whatever it is, would be at least half-way to beauty. Maybe all the way. We shall indeed be surprised at the result of attempting to give plain answers to questions such as these. Beauty will burst upon us unawares.

A SHORT LIST OF BOOKS AND OTHER PUBLICATIONS ON DESIGN

The books and papers included in this list amplify or bring up to date the ten chapters of *Design in Modern Life*. They are not highly technical works; all are easy to read, and they represent various points of view.

ARCHITECTURE

(a) GENERAL

Dr. E. Frankland Armstrong, F.R.S.
"Post-War Building—the Chemist's Contribution," a paper read before a joint meeting of the Road and Building Materials Group and the London Section of the Society and the Institution of Structural Engineers, on September 23, 1943. and published in *Chemistry and Industry*, February 12, 1944.

Christian Barman
Architecture (Benn's Sixpenny Library).
Balbus, or the Future of Architecture (Kegan Paul).

John Betjeman
Ghastly Good Taste (Chapman & Hall).

Darcy Braddell
How to Look at Buildings (Methuen).

Robert Byron
The Appreciation of Architecture (Wishart & Co).

W. A. Eden
The Process of Architectural Tradition (Macmillan).

A. Trystan Edwards
Architectural Style (Faber & Faber).
Good and Bad Manners in Architecture (Philip Allan).
The Things Which Are Seen (Philip Allan).

Maxwell Fry
Fine Building (Faber & Faber).

Frederick Gibberd
The Architecture of England (Architectural Press).

John Gloag
Men and Buildings (Country Life Ltd.).

W. H. Godfrey
Our Building Inheritance (Faber & Faber).

Walter Gropius
The New Architecture and the Bauhaus (Faber & Faber).

Henry Russell Hitchcock
In the Nature of Materials: 1887-1941 The Buildings of Frank Lloyd Wright (Duell, Sloane & Pearce, New York).

Julian Leathart
Style in Architecture (Nelson).

Le Corbusier
Towards a New Architecture (John Rodker).

WILLIAM LESCAZE
  On Being an Architect (G. P. Putnam's Sons, New York).
W. R. LETHABY
  Architecture (Home University Library).
NIKOLAUS PEVSNER
  An Outline of European Architecture (Penguin Books).
J. M. RICHARDS
  An Introduction to Modern Architecture (Penguin Books).
A. E. RICHARDSON and HECTOR O. CORFIATO
  The Art of Architecture (English Universities Press Ltd.).
HOWARD ROBERTSON
  Architecture Arising (Faber & Faber).
  Architecture Explained (Benn).
LOUIS H. SULLIVAN
  The Autobiography of an Idea (W. W. Norton & Company Inc., New York).
M. HARTLAND THOMAS
  "The Influence of New Developments in Construction on Architectural Design," a paper read at a meeting arranged by the Architectural Science Board on February 12, 1944, and published in the *Journal of the R.I.B.A.*, March, 1944.
DORA WARE and BETTY BEATTY
  A Short Dictionary of Architecture (Allen & Unwin).
C. AND A. WILLIAMS-ELLIS
  The Pleasures of Architecture (Cape: Life and Letters series).
CLOUGH WILLIAMS-ELLIS and JOHN SUMMERSON
  Architecture Here and Now (Nelson).
FRANK LLOYD WRIGHT
  Modern Architecture (the Kahn lectures for 1930) (Princeton University Press, U.S.A.).
F. R. S. YORKE and COLIN PENN
  A Key to Modern Architecture (Blackie).

(b) DOMESTIC

SIR PATRICK ABERCROMBIE
  The Book of the Modern House (Hodder & Stoughton).
SIDNEY O. ADDY
  Evolution of the English House (revised and enlarged from the Author's notes, by John Summerson) (Allen & Unwin).
CATHERINE BAUER
  Modern Housing (Allen & Unwin).
ANTHONY BERTRAM
  The House a machine for living in (A. & C. Black).
GEOFFREY BOUMPHREY
  Your House and Mine (Allen & Unwin).
HUGH BRAUN
  The Story of the English House (Batsford).

ELIZABETH DENBY
  Europe Re-housed (Allen & Unwin).
JOHN GLOAG
  The Englishman's Castle (Eyre & Spottiswoode).
NATHANIEL LLOYD
  A History of the English House (Architectural Press).
RAYMOND MCGRATH
  Twentieth Century Houses (Faber & Faber).
R.I.B.A. RECONSTRUCTION COMMITTEE
  Rebuilding Britain (Lund Humphries).
J. M. RICHARDS
  A Miniature History of the English House (Architectural Press).
F. R. S. YORKE and FREDERICK GIBBERD
  The Modern Flat (Architectural Press).

(c) PUBLIC BUILDINGS

H. P. ADAMS
  English Hospital Planning (R.I.B.A. Paper, 1929).
FELIX CLAY
  Modern School Buildings (Batsford).
L.C.C. REPORT
  Open Air Schools (King).
A. S. MELOY
  Theatres and Motion Picture Houses (Batsford).
L. G. PEARSON
  Recent Developments in Hospital Planning Abroad (R.I.B.A. paper, 1927).
P. MORTON SHAND
  Modern Theatres and Cinemas (Batsford).

TOWN AND COUNTRY PLANNING

T. ADAMS
  Recent Advances in Town Planning (Churchill).
S. D. ADSHEAD
  A New England: Planning for the Future (Muller).
ARCHITECTURAL PRESS
  Your Inheritance: An Uncomic Strip.
JOHN BETJEMAN
  English Cities and Small Towns (Collins: Britain in Pictures series).
GEOFFREY BOUMPHREY
  Town and Country Tomorrow (Nelson).
J. H. FORSHAW and PATRICK ABERCROMBIE
  County of London Plan (1943: Macmillan).
PATRICK GEDDES
  Cities in Evolution (Thornton Butterworth).

DESIGN IN MODERN LIFE

F. HAVERFIELD
Ancient Town Planning (Architectural Press).

JULIAN HUXLEY
TVA: Adventure in Planning (Architectural Press).

LE CORBUSIER
The City of Tomorrow and its Planning (John Rodker).

E. G. and G. McALLISTER
Town and Country Planning (Faber & Faber).

LEWIS MUMFORD
The Culture of Cities (Secker & Warburg).

FRANK PICK
Britain Must Rebuild (Kegan Paul).

C. B. PURDOM
Town Theory and Practice (Benn).

STEEN EILER RASMUSSEN
London: the Unique City (Cape).

ROYAL ACADEMY PLANNING COMMITTEE
Road, Rail and River in London: The second report of the Committee, with a Foreword by Sir Giles Gilbert Scott, R.A. July 1944. (Country Life Ltd.).

THOMAS SHARP
Town and Countryside (Oxford University Press).
Town Planning (Penguin Books).

LOUIS DE SOISSONS and ARTHUR KENYON
Site Planning in Practice at Welwyn Garden City (Benn).

RALPH TUBBS
Living in Cities (Penguin Books).

CLOUGH WILLIAMS-ELLIS
England and the Octopus (Geoffrey Bles).
Britain and the Beast (Dent).

HOME EQUIPMENT

GEOFFREY BOUMPHREY
The House—Inside and Out (Allen & Unwin).

NOEL CARRINGTON
Design in the Home (Country Life Ltd.).

JOHN GLOAG
English Furniture (A. & C. Black: Library of English Art).
Modern Home Furnishing (Macmillan's Sixpenny Library).

PAUL NASH
Room and Book (Soncino Press).

R. RANDALL PHILLIPS
The Servantless House (Country Life Ltd.).

M. and C. H. B. QUENNELL
A History of Everyday Things in England (Batsford).

N. C. Reynolds
Easier Housework (Country Life Ltd.).

T. H. Robsjohn-Gibbings
Good-bye, Mr. Chippendale (Alfred Knopf, New York).

John C. Rogers
Furniture and Furnishing (Oxford University Press).

INDUSTRIAL DESIGN AND MATERIALS

Anonymous
"Prime Cost Bows to Design," an article published in *Plastics*, June, 1941.

Dr. E. Frankland Armstrong, F.R.S.
"Materials, Old and New," a paper read before the Royal Society of Arts, on January 14, 1942, and published in the *Journal of the R.S.A.* No. 4608, Vol. XC, March 6, 1942.

Christian Barman
"Glass for Special Purposes," an article published in the *Official Architect*, June, 1944.

Geoffrey Boumphrey
The Story of the Ship (A. & C. Black).
The Story of the Wheel (A. & C. Black).

Sir William Bragg
Craftsmanship and Science (Watts & Co.).

Noel Carrington
Design and a Changing Civilisation (John Lane).

Metius Chappell
British Engineers (Collins: Britain in Pictures series).

Design and Industries Association
Four Lectures on Design, delivered by Henry Strauss, M.P., Francis Meynell, Tom Harrisson and Herbert Read, before the D.I.A., and ultimately issued in one publication (Hutchinson).

Sigfried Giedion
Space, Time and Architecture (Harvard University Press, U.S.A.; Oxford University Press).

John Gloag
Industrial Art Explained (Allen & Unwin).
The Missing Technician in Industrial Production (Allen & Unwin).
Plastics and Industrial Design, with a Section on the properties and uses of the various types of plastics by Grace Lovat Fraser (Allen & Unwin).
The Place of Glass in Building (Edited). (Allen & Unwin).
"The Influence of Plastics on Design," a paper read before the Royal Society of Arts on May 26, 1943, and published in the *Journal of the R.S.A.* No. 4644, Vol. XCI, July 23, 1943.
"Design for Industry," a paper read before a meeting of the Artists International Association, and published in the *Architects' Journal*, No. 2584, Vol. 100, August 3, 1944.

E. C. Goldsworthy
"Light Alloys in Post-War Britain," a paper read before the Royal Society of Arts on February 2, 1944, and published in the *Journal of the R.S.A.*, No. 4663, Vol. XCII, April 14, 1944.

Geoffrey Holme
Industrial Design and the Future (Studio Publications).

Raymond Loewy
"Selling through Design" a paper read for Mr. Loewy by Mr. John Gloag before the Royal Society of Arts on December 3, 1941, and published in the *Journal of the R.S.A.*, No. 2604, Vol. XC, January 9, 1942.

Raymond McGrath and A. C. Frost
Glass in Architecture and Decoration: with a Section on the nature and properties of Glass by H. E. Beckett (Architectural Press).

Lewis Mumford
Technics and Civilisation (Routledge).

"Plastes"
Plastics in Industry (Chapman & Hall).

Nikolaus Pevsner
Industrial Art in England (Cambridge University Press).
Pioneers of the Modern Movement: (From William Morris to Walter Gropius) (Faber & Faber).

Herbert Read
Art and Industry (Faber & Faber), 2nd edition 1944.

R. H. Sheppard
"Strength through Glass," an article published in the *Official Architect*, June, 1944.

Professor W. E. S. Turner
"New Uses for Glass," a paper read before the Royal Society of Arts on January 20, 1943, and published in the *Journal of the R.S.A.*, No. 4636, Vol. XCI, April 2, 1943.

John de la Valette (editor)
The Conquest of Ugliness (Methuen).

John W. Waterer
"The Industrial Designer and Leather," a paper read before the Royal Society of Arts on November 25, 1942, and published in the *Journal of the R.S.A.*, No. 4629. Vol. XCI, December 25, 1942.

V. E. Yarsley and E. G. Couzens
Plastics (Penguin Books), 3rd edition 1944.

CLOTHES

C. H. Ashdown
British Costume (Civil and Ecclesiastical) during Nineteen Centuries (Jack).

J. C. Flügel
The Psychology of Clothes (Hogarth Press).

Eric Gill
Clothes (Cape).

H. Hiler
From Nudity to Raiment (Foyle).

F. M. Kelly
Historic Costume (Batsford).

C. Köhler
The History of Costume (Harrap).

James Laver
Taste and Fashion from the French Revolution until To-day (Harrap).
"Fashion and War," a paper read before the Royal Society of Arts on March 1, 1944, and published in the *Journal of the R.S.A.*, No. 4666, Vol. XCII, May 26, 1944.

GENERAL

Anonymous
"Government and the Arts: the case for a United Front"; a leading article published in the *Times Literary Supplement*, No. 1842, May 22, 1937.

Clive Bell
Civilisation (Chatto & Windus).

Anthony Bertram
Design (Penguin Books).
Design in Daily Life (Methuen).

Sir Reginald Blomfield
Modernismus (Macmillan).

Margaret Bulley
Art and Counterfeit (Methuen).

Felix Clay
The Origin of the Sense of Beauty (Smith, Elder & Co.).

John Hemming Fry
The Revolt against Beauty (G. P. Putnam's Sons, New York).

Roger Fry
Vision and Design (Chatto & Windus).

Eric Gill
Beauty Looks after Herself (Sheed & Ward).

John Gloag (editor)
Design in Everyday Life and Things; the Year Book of the Design and Industries Association 1926–7 (Benn).

Milner Gray
"Civic Design," a paper read before the Design and Industrial Association at the Royal Society, and published in the *Architects' Journal*, September 14, 1944.

F. R. Leavis and Denys Thompson
Culture and Environment (Chatto & Windus).

W. R. Lethaby
Form in Civilisation (Oxford University Press).

DESIGN IN MODERN LIFE

LISLE MARCH PHILLIPPS
Form and Colour (Duckworth).
The Works of Man (Duckworth).

MANNING ROBERTSON
Everyday Architecture (Fisher Unwin).

SIR HUBERT LLEWELLYN SMITH
The Economic Laws of Art Production (Oxford University Press).

SIR HENRY TRUEMAN WOOD
A History of the Royal Society of Arts (John Murray).

SOME PERIODICALS CONCERNED WITH DESIGN

The Architectural Forum (New York) (monthly).
The Architectural Record (New York) (monthly).
The Architectural Review (monthly). This includes a regular feature reviewing industrial design.
Art and Industry (monthly).
The Studio (monthly).